Statistics and Measurement Concepts with OpenStat

William Miller

Statistics and Measurement Concepts with OpenStat

Springer

William Miller
Urbandale, Iowa
USA

ISBN 978-1-4614-5742-8   ISBN 978-1-4614-5743-5 (eBook)
DOI 10.1007/978-1-4614-5743-5
Springer New York Heidelberg Dordrecht London

Library of Congress Control Number: 2012952927

© Springer Science+Business Media New York 2013
This work is subject to copyright. All rights are reserved by the Publisher, whether the whole or part of the material is concerned, specifically the rights of translation, reprinting, reuse of illustrations, recitation, broadcasting, reproduction on microfilms or in any other physical way, and transmission or information storage and retrieval, electronic adaptation, computer software, or by similar or dissimilar methodology now known or hereafter developed. Exempted from this legal reservation are brief excerpts in connection with reviews or scholarly analysis or material supplied specifically for the purpose of being entered and executed on a computer system, for exclusive use by the purchaser of the work. Duplication of this publication or parts thereof is permitted only under the provisions of the Copyright Law of the Publisher's location, in its current version, and permission for use must always be obtained from Springer. Permissions for use may be obtained through RightsLink at the Copyright Clearance Center. Violations are liable to prosecution under the respective Copyright Law.
The use of general descriptive names, registered names, trademarks, service marks, etc. in this publication does not imply, even in the absence of a specific statement, that such names are exempt from the relevant protective laws and regulations and therefore free for general use.
While the advice and information in this book are believed to be true and accurate at the date of publication, neither the authors nor the editors nor the publisher can accept any legal responsibility for any errors or omissions that may be made. The publisher makes no warranty, express or implied, with respect to the material contained herein.

Printed on acid-free paper

Springer is part of Springer Science+Business Media (www.springer.com)

Preface

To the hundreds of graduate students and users of my statistics programs. Your encouragement, suggestions and patience have kept me motivated to maintain my interest in statistics and measurement.

To my wife who has endured my hours of time on the computer and wonders why I would want to create free material.

To my dogs Tuffy, Annie, Lacy and Heidi who sit under my desk for hours at a time while I pursue this hobby.

To Gary C. Ramseyer's First Internet Gallery of Statistics Jokes http://www.ilstu.edu/%7egcramsey/Gallery.html and Joachim Verhagen http://www.xs4all.nl/%7ejcdverha/scijokes

Urbandale, Iowa, USA   William G. Miller

Contents

1	**Basic Statistics**	1
	Introduction	1
	Symbols Used in Statistics	1
	Probability Concepts	4
	Additive Rules of Probability	5
	The Law of Large Numbers	6
	Multiplication Rule of Probability	7
	Permutations and Combinations	7
	Conditional Probability	8
	Bayesian Statistics	10
	Maximum Liklihood (Adapted from S. Purcell, http://statgen.iop.kcl.ac.uk/bgim/mle/sslike_1.html)	10
	A Simple Example of MLE	11
	Analytic MLE	13
	Numerical MLE	13
	Other Practical Considerations	14
	Log-Likelihood	15
	Model Identification	15
	Local Minima	16
	Probabilty as an Area	17
	Sampling	17
2	**Descriptive Statistics**	19
	The Mean	19
	Variance and Standard Deviation	22
	Estimating Population Parameters: Mean and Standard Deviation	23
	The Standard Error of the Mean	25
	Testing Hypotheses for Differences Between or Among Means	27
	The Nature of Scientific Investigation	27
	Decision Risks	28
	Hypotheses Related to a Single Mean	30

	Determining Type II Error and Power of the Test	33
	Sample Size Requirements for the Test of One Mean	36
	Confidence Intervals for a Sample Mean	39
	Frequency Distributions	40
	The Normal Distribution Model	42
	The Median	43
	Skew	44
	Kurtosis	44
	The Binomial Distribution	45
	The Poisson Distribution	46
	The Chi-Squared Distribution	47
	The F Ratio Distribution	48
	The "Student" t Test	49
3	**The Product Moment Correlation**	53
	Testing Hypotheses for Relationships Among Variables: Correlation	54
	Scattergrams	54
	Transformation to z Scores	56
	Simple Linear Regression	61
	The Least-Squares Fit Criterion	62
	The Variance of Predicted Scores	65
	The Variance of Errors of Prediction	66
	Testing Hypotheses Concerning the Pearson Product–Moment Correlation	67
	Hypotheses About Correlations in One Population	67
	Test That the Correlation Equals a Specific Value	68
	Testing Equality of Correlations in Two Populations	70
	Differences Between Correlations in Dependent Samples	72
	Partial and Semi-Partial Correlations	73
	Partial Correlation	74
	Semi-Partial Correlation	75
	Autocorrelation	76
	Series	84
	Introduction	84
4	**Multiple Regression**	87
	The Linear Regression Equation	87
	Least Squares Calculus	90
	Finding a Change in Y Given a Change in X for $Y = f(X)$	92
	Relative Change in Y for a Change in X	93
	The Concept of a Derivative	94
	Some Rules for Differentiating Polynomials	95
	Geometric Interpretation of a Derivative	97
	Finding the Value of X for Which $f(X)$ Is Least	98

A Generalization of the Last Example	99
Partial Derivatives	100
Least Squares Regression for Two or More Independent Variables	101
Matrix Form for Normal Equations Using Raw Scores	103
Matrix Form for Normal Equations Using Deviation Scores	104
Matrix Form for Normal Equations Using Standardized Scores	105
Hypothesis Testing in Multiple Regression	106
Testing the Significance of the Multiple Regression Coefficient	106
The Standard Error of Estimate	107
Testing the Regression Coefficients	107
Testing the Difference Between Regression Coefficients	109

5 Analysis of Variance 111

Theory of Analysis of Variance	111
The Completely Randomized Design	112
Introduction	112
A Graphic Representation	113
Null Hypothesis of the Design	113
Summary of Data Analysis	113
Model and Assumptions	114
Fixed and Random Effects	115
Analysis of Variance: The Two-Way, Fixed-Effects Design	115
Stating the Hypotheses	118
Interpreting Interactions	119
Random Effects Models	120
One Between, One Repeated Design	121
Introduction	121
The Research Design	121
Theoretical Model	123
Assumptions	123
Summary Table	124
Population Parameters Estimated	124
Two Factor Repeated Measures Analysis	125
Nested Factors Analysis of Variance Design	125
The Research Design	125
The Variance Model	126
The ANOVA Summary Table	126
A, B and C Factors with B Nested in A	126
Latin and Greco-Latin Square Designs	127
Some Theory	127
The Latin Square	128
Plan 1 by B. J. Winer	130
Plan 2	130

      Plan 3 Latin Squares Design 131
      Analysis of Greco-Latin Squares 132
      Plan 5 Latin Square Design 133
      Plan 6 Latin Squares Design 134
      Plan 7 for Latin Squares 134
      Plan 9 Latin Squares 135
  Analysis of Variance Using Multiple Regression Methods 136
      A Comparison of ANOVA and Regression 136
      Effect Coding 137
      Orthogonal Coding 138
      Dummy Coding 140
  Two Factor ANOVA by Multiple Regression 141
  Analysis of Covariance by Multiple Regression Analysis 144
  Sums of Squares by Regression 146
      The General Linear Model 146
  Canonical Correlation 147
      Introduction 147
      Eigenvalues and Eigenvectors 148
  The Canonical Analysis 151
  Discriminant Function/MANOVA 152
      Theory 152
  Cluster Analyses 153
      Theory 153
      Hierarchical Cluster Analysis 153
  Path Analysis 154
      Theory 154
  Factor Analysis 155
      The Linear Model 155

6 Non-Parametric Statistics 159
  Contingency Chi-Square 160
  Spearman Rank Correlation 160
  Mann–Whitney U Test 160
  Fisher's Exact Test 161
  Kendall's Coefficient of Concordance 161
  Kruskal-Wallis One-Way ANOVA 161
  Wilcoxon Matched-Pairs Signed Ranks Test 162
  Cochran Q Test 163
  Sign Test 163
  Friedman Two Way ANOVA 164
  Probability of a Binomial Event 164
  Runs Test 165
  Kendall's Tau and Partial Tau 166
  The Kaplan-Meier Survival Test 166
  Kolmogorov-Smirnov Test 166

7	**Statistical Process Control**	167
	Introduction	167
	XBAR Chart	167
	Range Chart	168
	S Control Chart	168
	CUSUM Chart	168
	p Chart	169
	Defect (Non-Conformity) c Chart	170
	Defects Per Unit u Chart	170
8	**Linear Programming**	171
	Introduction	171
	Calculation	172
	Implementation in Simplex	173
9	**Measurement**	175
	Test Theory	175
	Scales of Measurement	176
	Nominal Scales	176
	Ordinal Scales of Measurement	176
	Interval Scales of Measurement	177
	Ratio Scales of Measurement	178
	Reliability, Validity and Precision of Measurement	179
	Reliability	179
	The Kuder: Richardson Formula 20 Reliability	181
	Validity	184
	Concurrent Validity	185
	Predictive Validity	185
	Discriminate Validity	185
	Construct Validity	186
	Content Validity	187
	Effects of Test Length	188
	Composite Test Reliability	189
	Reliability by ANOVA	190
	Sources of Error: An Example	190
	A Hypothetical Situation	191
	Item and Test Analysis Procedures	199
	Classical Item Analysis Methods	200
	Item Discrimination	200
	Item Difficulty	201
	The Item Analysis Program	202
	Item Response Theory	203
	The One Parameter Logistic Model	205
	Estimating Parameters in the Rasch Model: Prox. Method	206

　　　　Item Banking and Individualized Testing 209
　　　　Measuring Attitudes, Values, Beliefs 210
　　　　Methods for Measuring Attitudes 211
　　　　　　Affective Measurement Theory 213
　　　　　　Thurstone Paired Comparison Scaling 214
　　　　　　Successive Interval Scaling Procedures 217
　　　　　　Guttman Scalogram Analysis 221
　　　　　　Likert Scaling 225
　　　　　　Semantic Differential Scales 226
　　　　　　Behavior Checklists 228
　　　　　　Codifying Personal Interactions 229

Bibliography 231

Index 235

List of Figures

Fig. 1.1	Maximum liklihood estimation	12
Fig. 1.2	Maximum liklihood estimation	16
Fig. 1.3	Local minima	17
Fig. 2.1	Distribution of sample means	32
Fig. 2.2	Sample size estimation	34
Fig. 2.3	Power curves	36
Fig. 2.4	Null and alternate hypotheses for sample means	37
Fig. 2.5	Sample plot of test scores	41
Fig. 2.6	Sample proportions of test scores	41
Fig. 2.7	Sample sumulative probabilities of test scores	42
Fig. 2.8	Sample probability plot	46
Fig. 2.9	A poisson distribution	47
Fig. 2.10	Chi-squared distribution with 4° of freedom	48
Fig. 2.11	t distribution with 2° of freedom	50
Fig. 2.12	t distribution with 100° of freedom	50
Fig. 3.1	A negative correlation plot	54
Fig. 3.2	Scattergram of two variables	55
Fig. 3.3	Scattergram of a negative relationship	55
Fig. 3.4	Scattergram of two variables with low relationship	56
Fig. 3.5	A simulated negative correlation plot	61
Fig. 3.6	X versus Y plot of five values	62
Fig. 3.7	Plot for a correlation of 1.0	63
Fig. 3.8	Single sample tests dialog form	70
Fig. 3.9	Form for comparison of correlations	74
Fig. 3.10	The autocorrelation dialog	78
Fig. 3.11	The moving average dialog	79
Fig. 3.12	Plot of smoothed points using moving averages	80
Fig. 3.13	Plot of residuals obtained using moving averages	80
Fig. 3.14	Polynomial regression smoothing form	81
Fig. 3.15	Plot of polynomial smoothed points	81

Fig. 3.16	Plot of residuals from polynomial smoothing	82
Fig. 3.17	Auto and partial autocorrelation plot	85
Fig. 4.1	A simple function map	91
Fig. 4.2	A function map in three dimensions	92
Fig. 4.3	The minimum of a function derivative	99

Chapter 1
Basic Statistics

> *It is proven that the celebration of birthdays is healthy. Statistics show that those people who celebrate the most birthdays become the oldest.*

Introduction

This chapter introduces the basic statistics concepts you will need throughout your use of the OpenStat package. You will be introduced to the symbols and formulas used to represent a number of concepts utilized in statistical inference, research design, measurement theory, multivariate analyses, etc. Like many people first starting to learn statistics, you may be easily overwhelmed by the symbols and formulas—don't worry, that is pretty natural and does NOT mean you are retarded! You may need to re-read sections several times however before a concept is grasped. You will not be able to read statistics like a novel (don't we wish we could) but rather must "study" a few lines at a time and be sure of your understanding before you proceed.

Symbols Used in Statistics

Greek symbols are used rather often in statistical literature. (Is that why statistics is Greek to so many people?) They are used to represent both arithmetic types of operations as well as numbers, called parameters, that characterize a population or larger set of numbers. The letters you usually use, called Arabic letters, are used for numbers that represent a sample of numbers obtained from the population of numbers.

Two operations that are particularly useful in the field of statistics that are represented by Greek symbols are the summation operator and the products operator. These two operations are represented by the capital Greek letters Sigma Σ and Pi Π. Whenever you see these symbols you must think:

$$\Sigma = \text{"The sum of the values :"}, \text{ or}$$

$$\Pi = \text{"The product of the values :"}$$

For example, if you see $Y = \Sigma (1,3,5,9)$ you would read this as "the sum of 1, 3, 5 and 9". Similarly, if you see $Y = \Pi(1,3,5,9)$ you would think "the product of 1 times 3 times 5 times 9".

Other conventions are sometimes adopted by statisticians. For example, as in beginning algebra classes, we often use X to represent any one of many possible numbers. Sometimes we use Y to represent a number that depends on one or more other numbers X1, X2, etc. Notice that we used subscripts of 1, 2, etc. to represent different (unknown) numbers. Lower case letters like y, x, etc. are also sometimes used to represent a deviation of a score from the mean of a set of scores. Where it adds to the understanding, X, and x may be italicized or written in a script style.

Now lets see how these symbols might be used to express some values. For example, we might represent the set of numbers (1,3,7,9,14,20) as X1, X2, X3, X4, X5, and X6. To represent the sum of the six numbers in the set we could write:

$$Y = \sum_{i=1}^{6} X_i = 1 + 3 + 7 + 9 + 14 + 20 = 54 \tag{1.1}$$

If we want to represent the sum of any arbitrary set of N numbers, we could write the above equation more generally, thus

$$Y = \sum_{i=1}^{N} X_i \tag{1.2}$$

represents the sum of a set of N values. Note that we read the above formula as "Y equals the sum of X subscript i values for the value of i ranging from 1 through N, the number of values".

What would be the result of the formula below if we used the same set of numbers (1,3,7,9,14,20) but each were multiplied by five ?

$$Y = \sum_{i=1}^{N} 5X_i = 5 \sum_{i=1}^{N} X_i = 270 \tag{1.3}$$

To answer the question we can expand the formula to

$$Y = 5X_1 + 5X_2 + 5X_3 + 5X_4 + 5X_5 + 5X_6$$
$$= 5(X_1 + X_2 + X_3 + X_4 + X_5 + X_6)$$
$$= 5(1 + 3 + 7 + 9 + 14 + 20)$$
$$= 5(54) = 270 \qquad (1.4)$$

In other words,

$$Y = \sum_{i=1}^{N} CX_i = C \sum_{i=1}^{N} X_i \qquad (1.5)$$

We may generalize multiplying any sum by a constant (C) to

$$Y = \sum_{i=1}^{N} CX_i = C \sum_{i=1}^{N} X_i \qquad (1.6)$$

What happens when we sum a term which is a compound expression instead of a simple value? For example, how would we interpret

$$Y = \sum_{i=1}^{N} (X_i - C) \qquad (1.7)$$

where C is a constant value?

We can expand the above formula as

$$Y = (X_1 - C) + (X_2 - C) + \ldots + (X_N - C) \qquad (1.8)$$

(Note the use of ... to denote continuation to the Nth term).
The above expansion could also be written as

$$Y = (X_1 + X_2 + \ldots + X_N) - NC \qquad (1.9)$$

$$\text{Or } Y = \sum_{i=1}^{N} X_i - NC \qquad (1.10)$$

We note that the sum of an expression which is itself a sum or difference of multiple terms is the sum of the individual terms of that expression. We may say that the summation operator distributes over the terms of the expression!

Now lets look at the sum of an expression which is squared. For example,

$$Y = \sum_{i=1}^{N} (X_i - C)^2 \qquad (1.11)$$

When the expression summed is not in its most simple form, we must first evaluate the expression. Thus

$$Y = \sum_{i=1}^{N} (X_i - C)^2 = \sum_{i=1}^{N} (X_i - C)(X_i - C) = \sum_{i=1}^{N} \left[X_i^2 - 2CX_i + C^2 \right]$$

$$= \sum_{i=1}^{N} X_i^2 - \sum_{i=1}^{N} 2CX_i + \sum_{i=1}^{N} C^2$$

$$\text{or } Y = \sum_{i=1}^{N} X_i^2 - 2C \sum_{i=1}^{N} X_i + NC^2 = \sum_{i=1}^{N} X^2 - 2CN\overline{X} - NC^2$$

$$= \sum_{i=1}^{N} X^2 - CN(2\overline{X} - C) \tag{1.12}$$

Probability Concepts

Maybe, possibly, could be, chances are, probably are all words or phrases we use to convey uncertainty about something. Yet all of these express some belief that a thing or event could occur or exist. The field of statistics is concerned about making such statements based on observations that will lead us to correct "guesses" about an event occuring or existing. The field of study called "statistics" gets its name from the use of samples that we can observe to estimate characteristics about the population that we cannot observe. If we can study the whole population of objects or events, there is no need for statistics! Accounting methods will suffice to describe the population. The characteristics (or indexes) we observe about a sample from a population are called *statistics*. These indexes are estimates of population characteristics called *parameters*. It is the job of the statistician to provide indexes (statistics) about populations that give us some level of *confidence* that we have captured the true characteristics of the population of interest.

When we use the term *probability* we are talking about the *proportion* of objects in some population. It might be the proportion of some discrete number of heads that we get when tossing a coin. It might be the proportion of values within a specific range of values we find when we observe test scores of student achievement examinations.

In order for the statistician to make useful observations about a sample that will help us make confident statements about the population, it is often necessary to make *assumptions* about the *distribution* of scores in the population. For example, in tossing a coin 30 times and examining the outcome as the number of heads or tails, the statistician would assume that the distribution of heads and tails after a very large number of tosses would follow the *binomial* distribution, a theoretical distribution of scores for a binary object. If the population of interest is the relationship between beginning salaries and school achievement, the statistician may have to assume that the measures of salary and achievement have a *normal* distribution and that the relationship can be described by the *bivariate-normal* distribution.

A variety of indexes (statistics) have been developed to estimate characteristics (measurements) of a population. There are statistics that describe the *central tendency* of the population such as the mean (average), median and mode. Other statistics are used to describe how variable the scores are. These statistics include the variance, standard deviation, range, semi-interquartile range, mean deviation, etc. Still other indices are used to describe the relationship among population characteristics (measures) such as the product–moment correlation and the multiple regression coefficient of determination. Some statistics are used to examine differences among samples from possibly different populations to see if they are more likely to be samples from the same population. These statistics include the "t" and "z" statistic, the chi-squared statistic and the F-Ratio statistic.

The sections below will describe many of the statistics obtained on samples to make inferences about population parameters. The assumed (theoretical) distribution of these statistics will also be described.

Additive Rules of Probability

Formal aspects of probability theory are discussed in this section. But first, we need to define some terms we will use. First, we will define a *sample space* as simply a set of points. A point can represent anything like persons, numbers, balls, accidents, etc. Next we define an *event*. An event is an observation of something happening such as the appearance of "heads" when a coin is tossed or the observation that a person you selected at random from a telephone book is voting Democrat in the next election. There may be several points in the sample space, each of which is an example of an event. For example, the sample space may consist of 5 black balls and 4 white balls in an urn. This sample space would have 9 points. An event might be "a ball is black." This event has 5 sample space points. Another event might be "a ball is white." This event has a sample space of 4 points. We may now say that the probability of an event E is the ratio of the number of sample points that are examples of E to the total number of sample points provided all sample points are equally likely. We will use the notation P(E) for the probability of an event. Now let an event be "A ball is black" where the sample space is the set of 9 balls (5 black and 4 white.) There are 5 sample points that are examples of this event out of a total of 9 sample points. Thus the probability of the event P(E) = 5/9. Notice that the probability that a ball is white is 4/9. We may also say that the probability that a ball is red is 0/9 or that the probability that the ball is both white and black is 0/9. What is the probability that the ball is either white OR black? Clearly this is (5 + 4)/9 = 1.0.

In our previous example of urn balls, we noticed that a ball is either white or black. These are mutually exclusive events. We also noted that the sum of exclusive events is 1.0. Now let us add 3 red balls to our urn. We will label our events as B, W or R for the colors they represent. Our sample space now has 12 points. What is the probability that two balls selected are either B or W? When the events are exclusive we may write this as P(B U A).

Since these are exclusive events, we can write: P(B U W) = P(B) + P(W) = 5/12 + 4/12 = 9/12 = 3/4 = 0.75.

It is possible for a sample point to be an example of two or more events. For example if we toss a "fair" coin three times, we can observe eight possible outcomes:

1. HHH 2. HHT 3. HTH 4. HTT 5. TTT 6. TTH 7. THT and 8. THH

If our coin is fair we can assume that each of these outcomes is equally likely, that is, has a probability of 1/8. Now let us define two events: event A will be getting a "heads" on flip 1 and flip 2 of the coin and event B will be getting a "heads" on flips 1 and 3 of the coin. Notice that outcomes 1 and 2 above are sample points of event A and that outcomes 1 and 3 are events of type B. Now we can define a new event that combines events A *and* B. We will use the symbol A ∩ B for this event. If we assume each of the eight sample points are equally likely we may write P(A ∩ B) = number of sample points that are examples of A ∩ B/total number of sample points, or

P(A ∩ B) = 1/8. Notice that only 1 of the points in our sample space has heads on both flips 1 and 2 and on 2 and 3 (sample point 1.) That is, the probability of event A *and* B is the probability that both events A and B occur.

When events may not be exclusive, we are dealing with the probability of an event A or Event B or both. We can then write

$$P(A \cup B) = P(A) + P(B) - P(A \cap B) \tag{1.13}$$

Which, in words says, the probability of events A or B equals the probability of event A plus the probability of event B minus the probability of event A and B. Of course, if A and B are mutually exclusive then the probabilty of A and B is zero and the probability of A or B is simply the sum of P(A) and P(B).

The Law of Large Numbers

Assume again that you have an urn of 5 black balls and 4 white balls. You stir the balls up and draw one from the urn and record its color. You return the ball to the urn, again stir the balls vigourously and again draw a single ball and record its color. Now assume you do this 10,000 times, each time recording the color of the ball. Finally, you count the number of white balls you drew from the 10,000 draws. You might reasonably expect the proportion of white balls to be close to 4/9 although it is likely that it is not exactly 4/9. Should you continue to repeat this experiment over and over, it is also reasonable to expect that eventually, the proportion would be extremely close to the actual proportion of 4/9. You can see that the larger the number of observations, the more closely we would approximate the actual value. You can also see that with very small replications, say 12 draws (with replacement) could lead to a very poor estimate of the actual proportion of white balls.

Multiplication Rule of Probability

Assume you toss a fair coin five times. What is the probability that you get a "heads" on all five tosses? First, the probability of the event $P(E) = 1/2$ since the sample space has only two possible outcomes. The multicative rule of probability states that the probability of five heads would be 1/2 * 1/2 * 1/2 * 1/2 * 1/2 or simply (1/2) to the fifth power (1/32) or, in general, $P(E)^n$ where n is the number of events E.

As another example of this rule, assume a student is taking a test consisting of six multiple-choice items. Each item has five equally attractive choices. Assume the student has absolutely no knowledge and therefore guesses the answer to each item by randomly selecting one of the five choices for each item. What is the probability that the student would get all of the items correct? Since each item has a probability of 1/5, the probability that all items are answered correctly is $(1/5)^6$ or 0.000064. What would it be if the items were true-false items?

Permutations and Combinations

A *permutation* is an arrangement of n objects. For example, consider the letters A, B, C and D. How many permutations (arrangements) can we make with these four letters? We notice there are four possibilities for the first letter. Once we have selected the first letter there are 3 possible choices for the second letter. Once the second letter is chosen there are two possibilities for the third letter. There is only one choice for the last letter. The number of permutations possible then is $4 \times 3 \times 2 \times 1 = 24$ ways to arrange the four letters. In general, if there are N objects, the number of permutations is $N \times (N-1) \times (N-2) \times (N-3) \times \ldots$ (1). We abbreviate this series of products with an exclamation point and write it simply as N! We say "N factorial" for the product series. Thus $4! = 24$. We do, however, have to let $0! = 1$, that is, by definition the factorial of zero is equal to one. Factorials can get very large. For example, $10! = 3,628,800$ arrangements. If you spent a minute examining one arrangement of 12 guests for a party, how long would it take you to examine each arrangement? I'm afraid that if you worked 8 hours a day, 5 days a week for 52 weeks a year you (and your descendants) would still be working on it for more than a 1,000 years!

A *combination* is a set of objects without regard to order. For example, the combination of A, B, C and D in any permutation is one combination. A question arises however concerning how many combinations of K objects can be obtained from a set of N objects. For example, how many combinations of 2 objects can be obtained from a set of 4 objects. In our example, we have the possibilities of A + B, A + C, A + D, B + C, B + D and C + D or a total of 6 combinations. Notice that the combination AB is the same as BA because order is not considered. A formula may be written using permutations that gives us a general formula for combinations. It is

$$N! / [K! (N-K)!] \tag{1.14}$$

In our example then, the number of combinations of 2 things out of 4 is 4!/[2! (4−2)!] which might be written as

$$\frac{4 \times 3 \times 2 \times 1}{(2 \times 1) \times (2 \times 1)} = \frac{24}{4} = 6 \tag{1.15}$$

A special mathematics notation is often used for the combination of k things out of N things. It is

$$\binom{N}{K} = \frac{N!}{K!(N-K)!} \tag{1.16}$$

You will see the use of combinations in the section on the binomial distribution.

Conditional Probability

In sections above we defined the additive law for mutually exclusive events as the sum of the invidual probabilities. For example, for a fair die the probability of each of the faces is 1/6 so the probability of getting a 1 in two tosses (toss A and a toss B) is $P(A) + P(B) = 1/6 + 1/6 = 1/3$. Our multiplicative law for independent events states that the probability of obtaining event A *and* event B is $P(A) \times P(B)$. So the probability of getting a 1 on toss A of a die 1 *and* toss B of the die is $P(1) \times P(2) = 1/6 \times 1/6 = 1/36$. But what if we don't know our die is a "fair" die with equal probabilties for each face on a toss? Can we use the prior information from toss A of the die to say what the probability if for toss B?

Conditional probability is the probability of an event given that another event has already occurred. We would write

$$P(B|A) = \frac{P(A \cap B)}{P(A)} \tag{1.17}$$

Conditional Probability

If A and B are independent then

$$P(B|A) = \frac{P(A)P(B)}{P(A)} = P(B) \tag{1.18}$$

or the probability of the second toss is 1/6, the same as before.

Now consider two events A and B: for B an individual has tossed a die four times with outcomes E1, E2, E3 and E4; For A the event is the tosses with outcomes E1 and E2. The events might be the toss results of 1, 3, 5 and 6. Knowing that event A has occurred, what is the probabilty of event B, that is, P(A|B)? Intuitively you might notice that the probabilty of the B event is the sum of the individual probabilities or 1/6 + 1/6 + 1/6 + 1/6 = 2/3, and that the probability of the A event is 1/6 + 1/6 = 1/3 or half the probability of B. That is, P(A)/P(B) = 1/2.

A more formal statement of conditional probability is

$$P(A|B) = \frac{P(A \cap B)}{P(B)} \tag{1.19}$$

Thus the probability of event A is conditional on the prior probability of B. The result P(A|B) is sometimes called the posterior probability. Notice we can rewrite the above equation as:

$$P(A|B)P(B) = P(A \cap B) \tag{1.20}$$

and

$$P(B|A)P(A) = P(A \cap B) \tag{1.21}$$

Since both equations equal the same thing we may write

$$P(A|B) = \frac{P(B|A)P(A)}{P(B)} \tag{1.22}$$

The above is known as Bayes Theorem for events.

Now consider an example. In a recent poll in your city, 40% are registered Democrats and 60% are registered Republicans. Among the Democrats, the poll shows that 70% feel that invading Iraq was a mistake and 20% feel it was justified. You have just met a new neighbor and have begun a conversation over a cup of coffee. You learn that this neighbor feels that invading Iraq was a mistake. What is the probability that the neighbor is also a Democrat? Let A be the event that the neighbor is Democrat and B be the event that she feels the invasion was a mistake. We already know that the probability of A is P(A) = 0.6. We also know that the probability of B is P(B|A) = 0.7 . We need to compute P(B), the probability the neighbor feels the invasion was a mistake. We notice that the probability of B can be decomposed into two exclusive parts: P(B) = P(B and A) and P(B and *not* A) where the probability of *not* A is 1—P(A) or 0.4, the probability of not being a democrat. We can write

$$P(B \cap notA) = P(notA)P(B|A) \tag{1.23}$$

$$\text{or } P(B) = P(B \text{ and } A) + P(notA)P(B|notA) \tag{1.24}$$

$$\text{or } P(B) = P(B|A)P(A) + P(notA)\,P(B|notA) \tag{1.25}$$

Now we know $P(A) = 0.4$, $P(not\ A) = 1-.4 = 0.6$, $P(B|A) = 0.7$ and $P(B|\ not\ A) = 0.2$. Therefore,

$$P(B) = (0.7)(0.4) + (0.6)(0.2) = 0.40$$

Now knowing P(B) we can compute P(A|B) using Bayes' Theorem:

$$P(A|B) = \frac{P(B|A)P(A)}{P(B)} = \frac{(0.7)(0.4)}{0.4} = 0.7 \tag{1.26}$$

is the probability of the neighbor being Democrat.

Bayesian Statistics

In the previous section we explored Bayes Theorem. In that discussion we had prior information P(A) and sought posterior probabilities of A given that B occurred. In general, Bayesian statistics follows this core:

> Prior Probabilities, e.g. $P(A)$ + New Information,
> e.g. apposssed to invading Iraq $P(B)$ = Posterior
> Probability $P(A|B)$.

The above example dealt with specific events. However, Bayesian statistics also can be generalized to situations where we wish to develop a posterior distribution by combining a prior distribution with a distribution of new information. The Beta distribution is often used for prior and posterior distributions. This text will not attempt to cover Bayesian statistics. The reader is encouraged to find text books specific to this topic.

Maximum Liklihood (Adapted from S. Purcell, http://statgen.iop.kcl.ac.uk/bgim/mle/sslike_1.html)

Model-Fitting

If the probability of an event X dependent on model parameters p is written

$$P(X\mid p)$$

then we would talk about the likelihood

$$L(p \mid X)$$

that is, the likelihood of *the parameters given the data*.

For most sensible models, we will find that certain data are more probable than other data. The aim of maximum likelihood estimation is to find the parameter value(s) that makes the observed data most likely. This is because the likelihood of the parameters given the data is defined to be equal to the probability of the data given the parameters (nb. technically, they are proportional to each other, but this does not affect the principle).

If we were in the business of making predictions based on a set of solid assumptions, then we would be interested in probabilities—the probability of certain outcomes occurring or not occurring.

However, in the case of *data analysis*, we have already observed all the data: once they have been observed they are fixed, there is no 'probabilistic' part to them anymore (the word data comes from the Latin word meaning 'given'). We are much more interested in the likelihood of the model parameters that underly the fixed data.

Probability
```
    Knowing parameters -> Prediction of outcome
```
Likelihood
```
    Observation of data -> Estimation of parameters
```

A Simple Example of MLE

To re-iterate, the simple principle of maximum likelihood parameter estimation is this: find the parameter values that make the observed data most likely. How would we go about this in a simple coin toss experiment? That is, rather than assume that p is a certain value (0.5) we might wish to find the *maximum likelihood estimate* (MLE) of p, given a specific dataset.

Beyond parameter estimation, the likelihood framework allows us to make *tests* of parameter values. For example, we might want to ask whether or not the estimated p differs *significantly* from 0.5 or not. This test is essentially asking: is there evidence that the coin is biased? We will see how such tests can be performed when we introduce the concept of a *likelihood ratio test* below.

Say we toss a coin 100 times and observe 56 heads and 44 tails. Instead of assuming that p is 0.5, we want to find the MLE for p. Then we want to ask whether or not this value differs significantly from 0.50.

How do we do this? We find the value for p that makes the observed data most likely.

As mentioned, the observed data are now fixed. They will be constants that are plugged into our binomial probability model :-

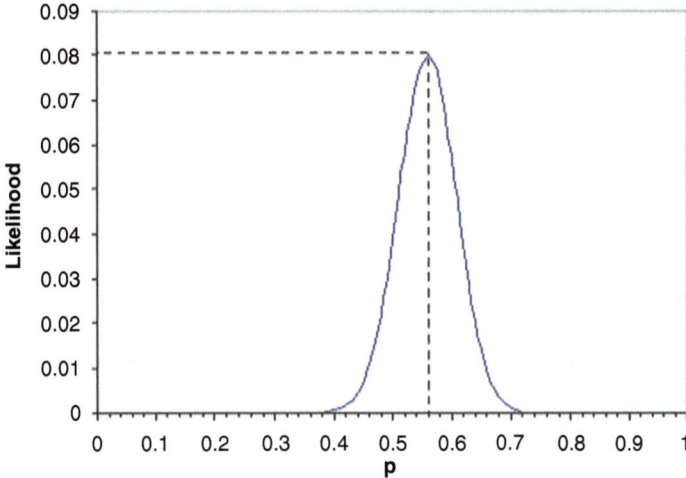

Fig. 1.1 Maximum liklihood estimation

- n = 100 (total number of tosses)
- h = 56 (total number of heads)

Imagine that p was 0.5. Plugging this value into our probability model as follows:

$$L(p = 0.5 | data) = \frac{100!}{56!44!} 0.5^{56} 0.5^{44} = 0.0389 \qquad (1.27)$$

But what if p was 0.52 instead?

$$L(p = 0.52 | data) = \frac{100!}{56!44!} 0.56^{56} 0.48^{44} = 0.0581 \qquad (1.28)$$

So from this we can conclude that p is more likely to be 0.52 than 0.5. We can tabulate the likelihood for different parameter values to find the maximum likelihood estimate of p:

```
  p         L
---------------
 0.48     0.0222
 0.50     0.0389
 0.52     0.0581
 0.54     0.0739
 0.56     0.0801
 0.58     0.0738
 0.60     0.0576
 0.62     0.0378
```

If we graph these data across the full range of possible values for p we see the following *likelihood surface* (Fig. 1.1).

Bayesian Statistics

We see that the maximum likelihood estimate for p seems to be around 0.56. In fact, it is exactly 0.56, and it is easy to see why this makes sense in this trivial example. The best estimate for p from any one sample is clearly going to be the proportion of heads observed in that sample. (In a similar way, the best estimate for the population mean will always be the sample mean.)

So why did we waste our time with the maximum likelihood method? In such a simple case as this, nobody would use maximum likelihood estimation to evaluate p. But not all problems are this simple! As we shall see, the more complex the model and the greater the number of parameters, it often becomes very difficult to make even reasonable guesses at the MLEs. The likelihood framework conceptually takes all of this in its stride, however, and this is what makes it the work-horse of many modern statistical methods.

Analytic MLE

Sometimes we can write a simple equation that describes the *likelihood surface* (e.g. the line we plotted in the coin tossing example) that can be differentiated. In this case, we can find the maximum of this curve by setting the first derivative to zero. That is, this represents the peak of a curve, where the gradient of the curve turns from being positive to negative (going left to right). In theory, this will represent the maximum likelihood estimate of the parameter.

Numerical MLE

But often we cannot, or choose not, to write an equation that can be differentiated to find the MLE parameter estimates. This is especially likely if the model is complex and involves many parameters and/or complex probability functions (e.g. the normal probability distribution).

In this scenario, it is also typically not feasible to evaluate the likelihood at all points, or even a reasonable number of points, in the *parameter space* of the problem as we did in the coin toss example. In that example, the parameter space was only one-dimensional (i.e. only one parameter) and ranged between 0 and 1. Nonetheless, because p can theoretically take any value between 0 and 1, the MLE will always be an approximation (albeit an incredibly accurate one) if we just evaluate the likelihood for a finite number of parameter values. For example, we chose to evaluate the likelihood at steps of 0.02. But we could have chosen steps of 0.01, of 0.001, of 0.000000001, etc. In theory and practice, one has to set a minimum *tolerance* by which you are happy for your estimates to be out. This is why computers are essential for these types of problems: they can tabulate lots and lots of values very quickly and therefore achieve a much finer resolution.

If the model has more than one parameter, the parameter space will grow very quickly indeed. Evaluating the likelihood exhaustively becomes virtually impossible—even for computers. This is why so-called *optimisation* (or *minimisation*) algorithms have become indispensable to statisticians and quantitative scientists in the last couple of decades. Simply put, the job of an optimisation algorithm is to *quickly* find the set of parameter values that make the observed data most likely. They can be thought of as intelligently playing some kind of hotter-colder game, looking for a hidden object, rather than just starting at one corner and exhaustively searching the room. The 'hotter-colder' information these algorithms utilise essentially comes from the way in which the likelihood *changes* as the they move across the parameter space. Note that it is precisely this type of 'rate of change' information that the analytic MLE methods use—differentiation is concerned with the rate of change of a quantity (i.e. the likelihood) with respect to some other factors (i.e. the parameters).

Other Practical Considerations

Briefly, we shall look at a couple of shortcuts and a couple of problems that crop up in maximum likelihood estimation using numerical methods:

Removing the Constant

Recall the likelihood function for the binomial distribution:

$$\frac{n!}{h!(n-h)!}p^h(1-p)^{n-h} \qquad (1.29)$$

In the context of MLE, we noted that the values representing the data will be fixed: these are n and h. In this case, the binomial 'co-efficient' depends only upon these constants. Because it does not depend on the value of the parameter p we can essentially ignore this first term. This is because any value for p which maximizes the above quantity will also maximize

$$\frac{n!}{h!(n-h)!} \qquad (1.30)$$

This means that the likelihood will have no meaningful scale in and of itself. This is not usually important, however, for as we shall see, we are generally interested not in the absolute value of the likelihood but rather in the *ratio* between two likelihoods—in the context of a likelihood ratio test.

We may often want to ignore the parts of the likelihood that do not depend upon the parameters in order to reduce the computational intensity of some problems. Even in the simple case of a binomial distribution, if the number of trials becomes very large, the calculation of the factorials can become infeasible (most pocket calculators cannot represent numbers larger than about 60!). (Note: in reality, we would quite probably use an approximation of the binomial distribution, using the normal distribution that does not involve the calculation of factorials).

Log-Likelihood

Another technique to make life a little easier is to work with the natural log of likelihoods rather than the likelihoods themselves. The main reason for this is, again, computational rather than theoretical. If you multiply lots of very small numbers together (say all less than 0.0001) then you will very quickly end up with a number that is too small to be represented by any calculator or computer as different from zero. This situation will often occur in calculating likelihoods, when we are often multiplying the probabilities of lots of rare but independent events together to calculate the joint probability.

With log-likelihoods, we simply add them together rather than multiply them (log-likelihoods will always be negative, and will just get larger (more negative) rather than approaching 0). Note that if

$$a = bc$$

then

$$log(a) = log(b) + log(c) \quad (1.31)$$

So, log-likelihoods are conceptually no different to normal likelihoods. When we optimize the log-likelihood (note: technically, we will be *minimizing* the *negative* log-likelihood) with respect to the model parameters, we also optimize the likelihood with respect to the same parameters, for there is a one-to-one (monotonic) relationship between numbers and their logs.

For the coin toss example above, we can also plot the log-likelihood. We can see that it gives a similar MLE for p (note: here we plot the negative of the log-likelihood, merely because most optimization procedures tend to be formulated in terms of minimization rather than maximization) (Fig. 1.2).

Model Identification

It is worth noting that it is not always possible to find one set of parameter values that uniquely optimises the log-likelihood. This may occur if there are too many

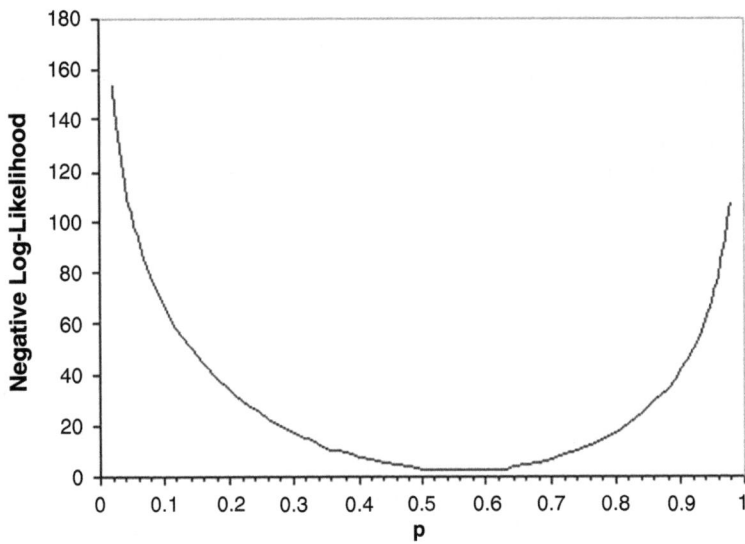

Fig. 1.2 Maximum liklihood estimation

parameters being estimated for the type of data that has been collected. Such a model is said to be 'under-identified'.

A model that attempted to estimate additive genetic variation, dominance genetic variation *and* the shared environmental component of variance from just MZ and DZ twin data would be under-identified.

Local Minima

Another common practical problem when implementing model-fitting procedures is that of local minima. Take the following graph, which represents the negative log-likelihood plotted by a parameter value, x (Fig. 1.3).

Model fitting is an iterative procedure: the user has to specify a set of *starting values* for the parameters (essentially an initial 'first guess') which the optimisation algorithm will take and try to improve on.

It is possible for the 'likelihood surface' to be any complex function of a parameter value, depending on the type of model and the data. In the case below, if the starting value for parameter x was at point A then optimisation might find the true, *global* minimum. However, if the starting value was at point B then it might not find instead only a local minimum. One can think of the algorithm crawling down the slope from B and thinking it has reached the lowest point when it starts to rise again. The implication of this would be that the optimisation algorithm would

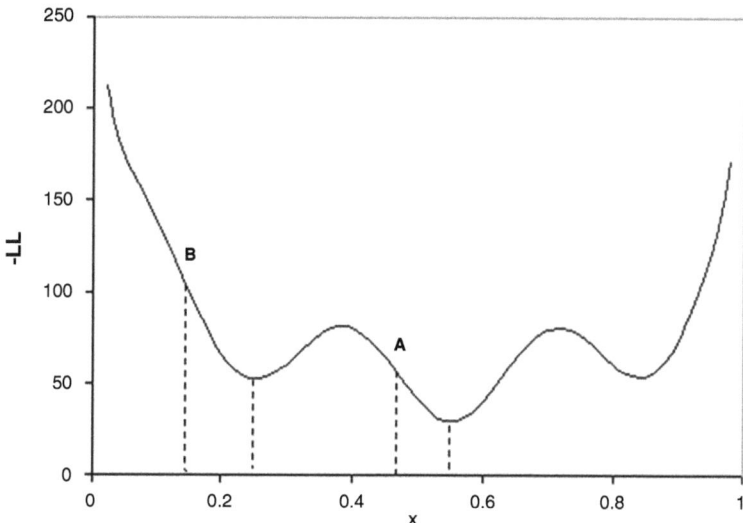

Fig. 1.3 Local minima

stop too early and return a sub-optimal estimate of the parameter x. Avoiding this kind of problem often involves specifying models well, choosing appropriate optimisation algorithms, choosing sensible starting values and more than a modicum of patience.

Probabilty as an Area

Probabilities are often represented as proportions of a circle or a polygon that shows the distribution of events in a sample space. Venn diagrams are circles with a portion of the ellipse shaded to represent a probability of an event in the space of the circle. In this case the circles area is considered to be 1.0. Distributions for binomial events, normally distributed events, poisson distributed events, etc. will often show a shaded area to represent a probability. You will see these shapes in sections to come.

Sampling

In order to make reasonable inferences about a population from a sample, we must insure that we are observing sample data that is not, in some artificial way, going to lead us to wrong conclusions about the population. For example, if we sample a group of Freshman college students about their acceptance or rejection of abortion, and use this to estimate the beliefs about the population of adults in the United States, we would not be collecting an *unbiased* or fair sample. We often use the term *experiment* to describe the process of drawing a sample. A *random experiment* or random sample is considered a fair or un-biased basis for estimating population parameters. You can appreciate the fact that the number of experiments (samples) drawn is highly critical to make relevant inferences about the population. For example, a series of four tosses of a coin and counting the number of heads that occur is a rather small number of samples from which to infer whether or not the coin is likely to yield 50% heads and 50% tails if you were to continue to toss the coin an infinite number of times! We will have much more confidence about our sample statistics if we use a large number of experiments.

Two of the most common mistakes of beginning researchers is failing to use a random sample and to use too few samples (observations) in their research. A third common mistake is to assume a theoretical model for the distribution of sample values that is incorrect for the population.

Chapter 2
Descriptive Statistics

The Mean

"When she told me I was average, she was just being mean".

The mean is probably the most often used parameter or statistic used to describe the central tendency of a population or sample. When we are discussing a population of scores, the mean of the population is denoted with the Greek letter μ. When we are discussing the mean of a sample, we utilize the letter X with a bar above it. The sample mean is obtained as

$$\bar{X} = \frac{\sum_{i=1}^{n} X_i}{n} \tag{2.1}$$

The population mean for a finite population of values may be written in a similar form as

$$\mu = \frac{\sum_{i=1}^{N} X_i}{N} \tag{2.2}$$

When the population contains an infinite number of values which are continuous, that is, can be any real value, then the population mean is the sum of the X values times the proportion of those values. The sum of values which can be an arbitrarily small in differences from one another is written using the integral symbol instead of the Greek sigma symbol. We would write the mean of a set of scores that range in size from minus infinity to plus infinity as

$$\mu = \int_{-\infty}^{+\infty} Xp(X)dx \qquad (2.3)$$

where p(X) is the proportion of any given X value in the population. The tall curve which resembles a script S is a symbol used in calculus to mean the "sum of" just like the symbol Σ that we saw previously. We use Σ to represent "countable" values, that is values which are discrete. The "integral" symbol on the other hand is used to represent the sum of values which can range continuously, that is, take on infinitely small differences from one-another.

A similar formula can be written for the sample mean, that is,

$$\overline{X} = \sum_{i=1}^{n} X_i p(X_i) \qquad (2.4)$$

where p(X) is the proportion of any given Xi value in the sample.

If a sample of n values is randomly selected from a population of values, the sample mean is said to be an unbiased estimate of the population mean. This simply means that if you were to repeatedly draw random samples of size n from the population, the average of all sample means would be equal to the population mean. Of course we rarely draw more than one or two samples from a population. The sample mean we obtain therefore will typically *not* equal the population mean but will in fact differ from the population mean by some specific amount. Since we usually don't know what the population mean is, we therefore don't know how far our sample mean is from the population mean. If we have, in fact, used random sampling though, we do know something about the shape of the distribution of sample means; they tend to be *normally* distributed. (See the discussion of the Normal Distribution in the section on Distributions). In fact, we can estimate how far the sample mean will be from the population mean some (P) percent of the time. The estimate of sampling errors of the mean will be further discussed in the section on testing hypotheses about the difference between sample means.

Now let us examine the calculation of a sample mean. Assume you have randomly selected a set of five scores from a very large population of scores and obtained the following:

$$X_1 = 3$$
$$X_2 = 7$$
$$X_3 = 2$$
$$X_4 = 8$$
$$X_5 = 5$$

The Mean

The sample mean is simply the sum (2.3) of the X scores divided by the number of the scores, that is

$$\bar{X} = \frac{\sum_{i=1}^{n} X_i}{n} = \sum_{i=1}^{5}(X_1 + X_2 + X_3 + X_4 + X_5)/5 = (3 + 7 + 2 + 8 + 5)/5 = 5.0 \quad (2.5)$$

We might also note that the proportion of each value of X is the same, that is, one out of five. The mean could also be obtained by

$$\bar{X} = \sum_{i=1}^{n} X_i p(X_i)$$
$$= 3\,(1/5) + 7\,(1/5) + 2\,(1/5) + 8\,(1/5) + 5\,(1/5) = 5.0 \quad (2.6)$$

The sample mean is used to indicate that value which is "most typical" of a set of scores, or which describes the center of the scores. In fact, in physics, the mean is the center-of-gravity (sometimes called the first moment of inertia) of a solid object and corresponds to the fulcrum, the point at where the object is balanced.

Unfortunately, when the population of scores from which we are sampling is not symmetrically distributed about the population mean, the arithmetic average is often not very descriptive of the "central" score or most representative score. For example, the population of working adults earn an annual salary of $21,000.00. These salaries however are not symmetrically distributed. Most people earn a rather modest income while there are a few who earn millions. The mean of such salaries would therefore not be very descriptive of the typical wage earner. The mean value would be much higher than most people earn. A better index of the "typical" wage earner would probably be the *median*, the value which corresponds to the salary earned by 50% or fewer people.

Examine the two sets of scores below. Notice that the first nine values are the same in both sets but that the tenth scores are quite different. Obtain the mean of each set and compare them. Also examine the score below which 50% of the remaining scores fall. Notice that it is the same in both sets and better represents the "typical" score.

SET A: (1, 2, 3, 4, 5, 6, 7, 8, 9, 10)

    Mean = ?
    Median = ?

SET B. (1, 2, 3, 4, 5, 6, 7, 8, 9, 1000)

    Mean = ?
    Median = ?

Did you know that the great majority of people have more than the average number of legs? It's obvious really; amongst the 57 million people in Britain there are probably 5,000 people who have got only one leg. Therefore the average number of legs is: ((5000 * 1) + (56,995,000 * 2)) / 57,000,000 = 1.9999123 Since most people have two legs...

Variance and Standard Deviation

A set of scores are seldom all exactly the same if they represent measures of some attribute that varies from person to person or object to object. Some sets of scores are much more variable that others. If the attribute measures are very similar for the group of subjects, then they are less variable than for another group in which the subjects vary a great deal. For example, suppose we measured the reading ability of a sample of 20 students in the third grade. Their scores would probably be much less variable than if we drew a sample of 20 subjects from across the grades 1 through 12!

There are several ways to describe the variability of a set of scores. A very simple method is to subtract the smallest score from the largest score. This is called the *exclusive range*. If we think the values obtained from our measurement process are really point estimates of a continuous variable, we may add 1 to the exclusive range and obtain the *inclusive range*. This range includes the range of possible values. Consider the set of scores below:

$$5, 6, 6, 7, 7, 7, 8, 8, 9$$

If the values represent discrete scores (not simply the closest value that the precision of our instrument gives) then we would use the exclusive range and report that the range is $(9-5) = 4$. If, on the other hand, we felt that the scores are really point estimates in the middle of intervals of width 1.0 (for example the score seven is actually an observation someplace between 6.5 and 7.5) then we would report the range as $(9-5) + 1 = 5$ or $(9.5-4.5) = 5$.

While the range is useful in describing roughly how the scores vary, it does not tell us much about how MOST of the scores vary around, say, the mean. If we are interested in how much the scores in our set of data tend to differ from the mean score, we could simply average the distance that each score is from the mean. The mean deviation, unfortunately is always 0.0! To see why, consider the above set of scores again:

$$\text{Mean} = (5+6+6+7+7+7+8+8+9) / 9 = 63 / 9 = 7.0$$

Now the deviation of each score from the mean is obtained by subtracting the mean from each score:

$$5 - 7 = -2$$
$$6 - 7 = -1$$
$$6 - 7 = -1$$
$$7 - 7 = 0$$
$$7 - 7 = 0$$
$$7 - 7 = 0$$
$$8 - 7 = +1$$
$$8 - 7 = +1$$
$$9 - 7 = +2$$
$$\text{Total} = 0.0$$

Since the sum of deviations around the mean always totals zero, then the obvious thing to do is either take the average of the absolute value of the deviations OR take the average of the squared deviations. We usually average the squared deviations from the mean because this index has some very important application in other areas of statistics.

The average of squared deviations about the mean is called the *variance* of the scores. For example, the variance, which we will denote as S^2, of the above set of scores would be:

$$S^2 = \frac{(-2)^2 + (-1)^2 + (-1)^2 + 0^2 + 0^2 + 0^2 + 1^2 + 1^2 + 2^2}{9} = 1.3333$$

approximately. (2.7)

Thus we can describe the score variability of the above scores by saying that the average squared deviation from the mean is about 1.3 score points.

We may also convert the average squared value to the scale of our original measurements by simply taking the square root of the variance, e.g. $S = \sqrt{(1.3)} = 1.1547$ (approximately). This index of variability is called the *standard deviation* of the scores. It is probably the most commonly used index to describe score variability!

Estimating Population Parameters: Mean and Standard Deviation

We have already seen that the mean of a sample of scores randomly drawn from a population of scores is an estimate of the population's mean. What we have to do is to imagine that we repeatedly draw samples of size n from our population (always placing the previous sample back into the population) and calculate a sample mean each time. The average of all (infinite number) of these sample means is the population mean. In algebraic symbols we would write:

$$\mu = \frac{\sum_{i=1}^{k} \overline{X}_i}{k} \quad \text{as } k \to \infty \tag{2.8}$$

Notice that we have let $\overline{X}$ represent the sample mean and μ represent the population mean. We say that the sample mean is an *unbiased* estimate of the population mean because the average of the sample statistic calculated in the same way that we would calculate the population mean leads to the population mean. We calculate the sample mean by dividing the sum of the scores by the number of scores. If we have a finite population, we could calculate the population mean in exactly the same way.

The sample variance calculated as the average of squared deviations about the sample mean is, however, a *biased* estimator of the population variance (and therefore the standard deviation also a biased estimate of the population standard deviation). In other words, if we calculate the average of a very large (infinite) number of sample variances this average will NOT equal the population variance. If, however, we multiply each sample variance by the constant $n/(n-1)$ then the average of these "corrected" sample variances will, in fact, equal the population variance! Notice that if n, our sample size, is large, then the bias $n/(n-1)$ is quite small. For example a sample size of 100 gives a correction factor of about 1.010101. The bias is therefore approximately one hundredth of the population variance. The reason that the average of squared deviations about the sample means is a biased estimate of the population variance is because we have a slightly different mean (the sample mean) in each sample.

If we had knowledge of the population mean μ and always subtracted μ from our sample values X, we would not have a biased statistic. Sometimes statisticians find it more convenient to use the biased estimate of the population variance than the unbiased estimate. To make sure we know which one is being used, we will use different symbols for the biased and unbiased estimates. The biased estimate will be represented here by a S^2 and the unbiased by a s^2. The reason for use of the square symbol is because the square root of the variance is the standard deviation. In other words we use S for the biased standard deviation and s for the unbiased standard deviation. The Greek symbol sigma σ is used to represent the population standard deviation and σ^2 represents the population variance. With these definitions in mind then, we can write:

$$\sigma^2 = \frac{\sum_{j=1}^{K} s_i^2}{k} \quad \text{as } k \to \infty \tag{2.9}$$

or

$$\sigma^2 = \frac{\sum_{j}^{k} \frac{n}{n-1} S_j^2}{k} \quad \text{as } k \to \infty \tag{2.10}$$

where n is the sample size, k the number of samples, S^2 is the biased sample variance and s^2 is the unbiased sample variance.

You may have already observed that multiplying the biased sample variance by n/(n−1) gives a more direct way to calculate the unbiased variance, that is:

$$s^2 = (n / (n-1)) * S^2$$

or

$$s^2 = \frac{n}{n-1} \frac{\sum_{i=1}^{n}(X_i - \overline{X})^2}{n} = \frac{\sum_{i=1}^{n}(X_i - \overline{X})^2}{n-1} \quad (2.11)$$

In other words, we may directly calculate the unbiased estimate of population variance by dividing the sum of square deviations about the mean by the sample size minus 1 instead of just the sample size.

The numerator term of the variance is usually just called the "sum of squares" as sort of an abbreviation for the sum of squared deviations about the mean. When you study the Analysis of Variance, you will see a much more extensive use of the sum of squares. In fact, it is even further abbreviated to SS . The unbiased variance may therefore be written simply as

$$s^2 = \frac{SS_x}{n-1}$$

The Standard Error of the Mean

In the previous discussion of unbiased estimators of population parameters, we discussed repeatedly drawing samples of size n from a population with replacement of the scores after drawing each sample. We noted that the sample mean would likely vary from sample to sample due simply to the variability of the scores randomly selected in each sample. The question may therefore be asked "How variable ARE the sample means?". Since we have already seen that the variance (and standard deviation) are useful indexes of score variability, why not use the same method for describing variability of sample means? In this case, of course, we are asking how much do the sample means tend to vary, on the average, around the population mean. To find our answer we could draw, say, several hundred samples of a given size and calculate the average of the sample means to estimate Since we have already seen that the variance (and standard deviation) are useful indexes of score variability, why not use the same method for describing variability of sample means? In this case, of course, we are asking how much do the sample means tend to vary, on the average, around the population mean. To find our answer we could draw, say, several hundred samples of a given size and calculate the average of the sample means to estimate: and then get the squared difference of each sample mean from this estimate. The average of these squared deviations would give us an approximate answer. Of course, because we did not draw ALL possible samples, we would still potentially have some error in our estimate. Statisticians have provided mathematical proofs of a more simple, and unbiased, estimate of how much the sample mean is expected to vary. To estimate the variance of sample means we simply draw ONE sample, calculate the unbiased estimate of X score variability in the population then divide that by the sample size! In symbols

$$s_{\bar{X}}^2 = \frac{s_X^2}{n} \tag{2.12}$$

The square root of this estimate of variance of sample means is the estimate of the standard deviation of sample means. We usually refer to this as the *standard error of the mean*. The standard error of the mean represents an estimate of how much the means obtained from samples of size n will tend to vary from sample to sample. As an example, let us assume we have drawn a sample of seven scores from a population of scores and obtained:

$$1, 3, 4, 6, 6, 2, 5$$

First, we obtain the sample mean and variance as:

$$\bar{X} = \frac{\sum_{i=1}^{7} X_i}{7} = 3.857 \text{ (approximately)} \tag{2.13}$$

$$s^2 = \frac{\sum_{i=1}^{7}(X_i - \bar{X})^2}{7 - 1} = \frac{127}{6} = 3.81 \tag{2.14}$$

Then the variance of sample means is simply

$$s_{\bar{X}}^2 = \frac{s_X^2}{n} = \frac{3.81}{7} = 0.544 \tag{2.15}$$

and the standard error of the mean is estimated as

$$s_{\bar{X}} = \sqrt{s_{\bar{X}}^2} = 0.74 \tag{2.16}$$

You may have noticed by now, that as long as we are estimating population parameters with sample statistics like the sample mean and sample standard deviation, that it is theoretically possible to obtain estimates of the variability of ANY sample statistic. In principle this is true, however, there are relatively few that have immediate practical use. We will only be using the expected variability of a few sample statistics. As we introduce them, we will tell you what the estimate is of the variance or standard deviation of the statistic. The standard error of the mean, which we just examined, will be used in the z and t-test statistic for testing hypotheses about single means. More on that later.

Testing Hypotheses for Differences Between or Among Means

The Nature of Scientific Investigation

People have been trying to understand the things they observe for as long as history has been recorded. Understanding observed phenomenon implies an ability to describe and predict the phenomenon. For example, ancient man sought to understand the relationship between the sun and the earth. When man is able to predict an occurrence or change in something he observes, it affords him a sense of safety and control over events. Religion, astrology, mysticism and other efforts have been used to understand what we observe. The scientific procedures adopted in the last several hundred years have made a large impact on human understanding. The scientific process utilizes inductive and deductive logic and the symbols of logic, mathematics. The process involves:

(a) Making systematic observations (**description**)
(b) Stating possible relationships between or differences among objects observed (**hypotheses**)
(c) Making observations under controlled or natural occurrences of the variations of the objects hypothesized to be related or different (**experimentation**)
(d) Applying an accepted decision rule for stating the truth or falsity of the speculations (**hypothesis testing**)
(e) Verifying the relationship, if observed (**prediction**)
(f) Applying knowledge of the relationship when verified (**control**)
(g) Conceptualizing the relationship in the context of other possible relationships (**theory**).

The rules for deciding the truth or falsity of a statement utilizes the assumptions developed concerning the chance occurrence of an event (observed relationship or difference). These decision rules are particularly acceptable because the user of the rules can ascertain, with some precision, the likelihood of making an error, whichever decision is made!

As an example of this process, consider a teacher who observes characteristics of children who mark false answers true in a true-false test as different from children who mark true answers as false. Perhaps the hypothetical teacher happens to notice that the proportion of left-handed children is greater in the first group than the second. Our teacher has made a systematic observation at this point. Next, the teacher might make a **scientific statement** such as "Being left-handed increases the likelihood of responding falsely to true-false test items." Another way of making this statement however could be "The proportion of left-handed children selecting false options of true statements in a true-false test does not differ from that of right handed children beyond that expected by sampling variability alone." This latter statement may be termed a **null hypothesis** because it states an absence (null) of a difference for the groups observed. The null hypothesis is the statement generally accepted for testing because the alternatives are innumerable. For example (1) no difference exists or (2) some difference exists. The scientific statement which states the principle of interest would be difficult to test because the possible differences are innumerable. For example, "increases" in the example above is not specific enough. Included in the set of possible "increases" are 0.0001, 0.003, 0.012, 0.12, 0.4, etc. After stating the null hypothesis, our scientist-teacher would make **controlled observations**. For example, the number of "false" options chosen by left and right handed children would be observed after controlling for the total number of items missed by each group. This might be done by **matching** left handed children with right handed children on the total test scores. The teacher may also need to insure that the number of boys and girls are also matched in each group to control for the possibility that sex is the variable related to option choices rather than handedness. We could continue to list other ways to control our observations in order to rule out variables other than the hypothesized ones possibly affecting our **decision**.

Once the teacher has made the controlled observations, decision rules are used to **accept or reject** the null hypothesis. We will discover these rules involve the chances of rejecting a true null hypothesis (**Type I error**) as well as the chances of accepting a false null hypothesis (**Type II error**).

Because of the chances of making errors in applying our decision rules, results should be verified through the observation of additional samples of subjects.

Decision Risks

Many research decisions have different losses which may be attached to outcomes of an experiment. The figure below summarizes the possible outcomes in testing a null hypothesis. Each outcome has a certain probability of occurrence. These probabilities (chances) of occurrence are symbolized by Greek letters in each outcome cell.

Possible Outcomes of an Experiment

		True State of Nature	
		H_o True	H_o False
Experimenter conclusion based on observed data	accept H_o	$1 - \alpha$	β Type II error
	reject H_o	Type I Error α	$1 - \beta$

In the above figure α (alpha) is the chance of obtaining a sample which leads to rejection of the null hypothesis when in the population from which the sample is drawn the null hypothesis is actually true. On the other hand, we also have the chance of drawing a sample that leads us to accept a null hypothesis when, in fact, in the population we should reject it. This latter error has ß (Beta) chances of occurring. Greek symbols have been used rather than numbers because the experimenter may control the types of error! For example, by selecting large samples, by reducing the standard deviation of the observed variable (for example by improving the precision of measurement), or by decreasing the size of the discrepancy (difference) we desire to be sensitive to, we can control both Type I and Type II error.

Typically, the chances of getting a Type I error is arbitrarily set by the researcher. For example, the value of alpha may be set to .05. Having set the value of α, the researcher can establish the sample size needed to control Type II error which is also arbitrarily chosen (e.g. ß = 0.2). In other cases, the experimenter is limited to the sample size available. In this case the experimenter must also determine the smallest difference or effect size (alternate hypothesis) to which he or she wishes to be sensitive.

How does a researcher decide on α, ß and a minimum discrepancy? By assessing or estimating the loss or consequences in making each type of error! For example, in testing two possible cancer treatments, consider that treatment 1 costs $1,000 while treatment 2 costs $100. Consider the null hypothesis

H_o: no difference between treatments (i.e. equally effective)

and consider the alternative

H_1: treatment 1 is more effective than treatment 2.

If we reject H_o: and thereby accept H_1: we will pay more for cancer treatment. We would probably be glad to do this if treatment 1 were, in fact, more effective. But if we have made a Type I error, our losses are 10 to 1 in dollars lost. On the other hand, consider the loss if we should accept H_o: when, in fact, H_1: is correct. In this case lives will be lost that might have been saved. What is one life worth? Most people would probably place more than $1,000 value on a life. If so, you would probably choose a smaller ß value than for α. The size of both these values are dependent on the size of risk you are willing to take. In the above example, a ß = 0.001 would not be unreasonable.

Part of our decision concerning α and ß also is based on the cost for obtaining each observation. Sometimes **destructive** observation is required. For example, in testing the effectiveness of a manufacturer's military missiles, the sample drawn would be destroyed by the testing. In these cases, the cost of additional observations may be as large as the losses associated with Type I or Type II error!

Finally, the size of the discrepancy selected as "meaningful" will affect costs and error rates. For example, is an IQ difference of five points between persons of Group A versus Group B a "practical" difference? How much more quickly can a child of 105 IQ learn over a child of 100 IQ? The larger the difference selected, the smaller is the sample needed to be sensitive to true population differences of that size. Thus, cost of data collection may be conserved by selecting realistic differences for the alternative hypothesis. If sample size is held constant while the discrepancy is increased, the chance of a Type II error is reduced, thus reducing the chances of a loss due to this type of error. We will examine the relationships between Type I and Type II error, the discrepancy chosen for an alternative hypothesis, and the sample size and variable's standard deviation in the following sections.

Hypotheses Related to a Single Mean

In order to illustrate the principles of hypothesis testing, we will select an example that is rather simple. Consider a hypothetical situation of the teacher who has administered a standardized achievement test in algebra to high school students completing their first course in algebra. Assume that extensive "norms" exist for the test showing that the population of previously tested students obtained a mean score equal to 50 and a standard deviation equal to 10. Further assume the teacher has 25 students in the class and that the class test mean was 55 and the standard deviation was 9. The teacher feels that his particular method of instruction is superior to those used by typical instructors and results in superior student performance. He wishes to provide evidence for his claim through use of the standardized algebra test. However, other algebra teachers in his school claim his teaching is really no better than theirs but requires half again as much time and effort. They would like to see evidence to substantiate their claim of no difference. What must our teachers do? The following steps are recommended by their school research consultant:

1. Agree among themselves how large a difference between the past population mean and the mean of the sampled population is a practical increment in algebra test performance.
2. Agree upon the size of Type I error they are willing to accept considering the consequences.
3. Because sample size is already fixed (n = 25), they cannot increase it to control Type II error. They can however estimate what it will be for the alternative

Hypotheses Related to a Single Mean

hypothesis that the sampled population mean does differ by a value as large or larger than that agreed upon in (2) above.

4. Use the results obtained by the classroom teacher to accept or reject the null hypothesis assuming that the sample means of the kind obtained by the teacher are normally distributed and unbiased estimates of the population mean. This is equivalent to saying we assume the teacher's class is a randomly selected sample from a population of possible students taught be the instructor's method. We also assume that the effect of the instructor is independent for each student, that is, that the students do not interact in such a way that the score of one student is somehow dependent on the score obtained by another student.

By assuming that sample means are normally distributed, we may use the probability distribution of the normally distributed z to test our hypothesis. Based on a theorem known as the "Central Limit Theorem", it can be demonstrated that sample means obtained from scores that are NOT normally distributed themselves DO tend to be normally distributed! The larger the sample sizes, the closer the distribution of sample means approaches the normal distribution. You may remember that our z score transformation is

$$z = \frac{X - \overline{X}}{S_x} = \frac{d}{S_x} \qquad (2.17)$$

when determining an individual's z score in a sample. Now consider our possible sample means in the above experiment to be individual scores that deviates (d) from a population mean (μ) and have a standard deviation equal to

$$S_{\overline{X}} = \frac{S_x}{\sqrt{n}} \qquad (2.18)$$

That is, the sample means vary inversely with the square root of the sample size. The standard deviation of sample means is also called the standard error of the mean. We can now transform our sample mean (55) into a z score where $\mu = 50$ and the standard error is $S_e = S_x/\sqrt{n} = 10/5 = 2$. Our result would be:

$$z_0 = \frac{\overline{X} - \mu_0}{S_e} = \frac{55 - 50}{2} = 2.5 \qquad (2.19)$$

Note we have used a small zero subscript by the population mean to indicate this is the null hypothesis mean.

Before we make any inference about our teacher's student performance, let us assume that the teachers agreed among themselves to set the risk of a Type I error rather low, at 0.05, because of the inherent loss of greater effort and time on their part if the hypothesis is rejected (assuming they adopt the superior teaching method). Let us also assume that the teachers have agreed that a class that achieves an average mean at least 2 standard deviations of the sample means above the previous

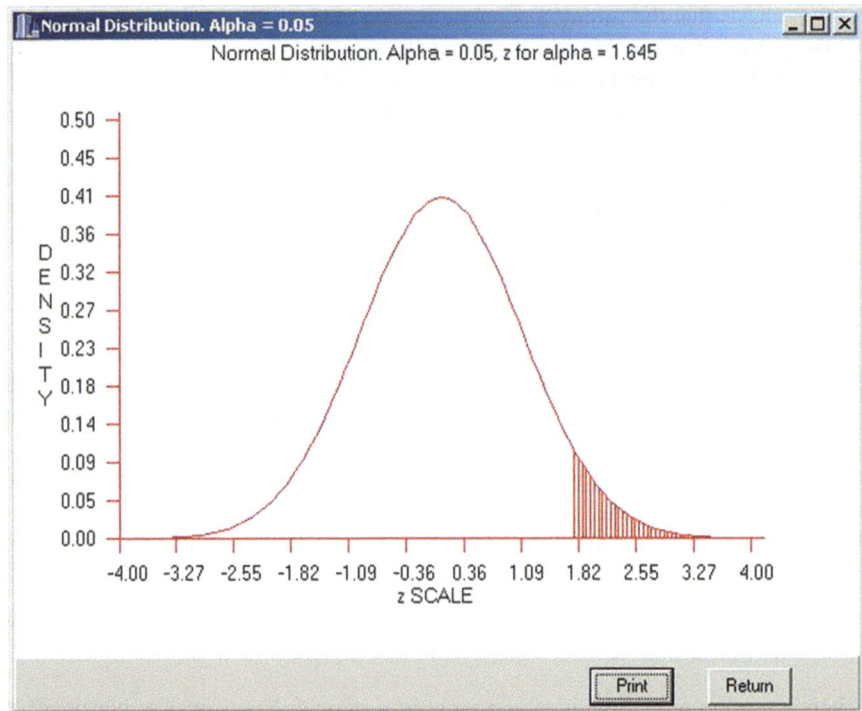

Fig. 2.1 Distribution of sample means

population mean is a realistic or practical increment in algebra learning. This means that the teachers want a difference of at least 4 points from the mean of 50 since the standard error of the means is 2.

Now examine the figure. In this figure the distribution of sample means is shown (since the statistic of interest is the sample mean.) A small caret (^) may be shown at the scale point where our specific sample statistic (the mean) falls in the theoretical distribution that has a mean of 50 and standard error of 2. Also shown, by shading is the area corresponding to the extreme 0.05 area of the distribution (Fig. 2.1).

Examination of the previous figure indicates that the sample mean obtained deviates from the hypothesized mean by a considerable amount (5 points). If we were obtaining samples from a population in which the mean was 50 and the standard error of the means was 2, we would expect to obtain a sample this deviant only 0.006 of the time! That is, only 0.006 of normally distributed z scores are as large or larger than the $z = 2.5$ that we obtained! Because our sample mean is *SO* deviant for the hypothesized population, we reject the hypothesized population mean and instead accept the alternative that the population from which we did sample has a mean greater than 50. If our statistic had not exceeded the z score corresponding to our Type I error rate, we would have accepted the null hypothesis.

Using a table of the normally distributed z score you can observe that the critical value for our decision is a $z_\alpha = 1.645$.

To summarize our example, we have thus far:

1. Stated our hypothesis. In terms of our critical z score corresponding to μ, we may write the hypothesis as

$$H_0 : z < z_\mu \qquad (2.20)$$

2. Stated our alternate hypothesis which is

$$H_1 : z > z_\mu$$

3. Obtained sample data and found that $z > z_\mu$ which leads us to reject H0: in favor of H_1:

Determining Type II Error and Power of the Test

In the example described above, the teachers had agreed that a deviation as large as 2 times the standard deviation of the means would be a "practical" teaching gain. The question may be asked, "What is the probability of accepting the null hypothesis when the true population mean is, in fact, 2 standard deviations (standard error) units above the hypothesized mean?" The figure below illustrates the theoretical distributions for both the null hypothesis and a specific alternate hypothesis, i.e. $H_1 = 54$ (Fig. 2.2).

The area to the left of the α value of 1.645 (frequently referred to as the region of rejection) under the null distribution (left-most curve) is the area of "acceptance" of the null hypothesis—any sample mean obtained that falls in this region would lead to acceptance of the null hypothesis. Of course, any sample mean obtained that is larger than the $z = 1.645$ would lead to rejection (the shaded portion of the null distribution). Now we may ask, "If we consider the alternative distribution (i.e. $\mu = 54$), what is the z value in that distribution which corresponds to the z value for μ under the null distribution?" To determine this value, we will first transform the z score for alpha under the null distribution back to the raw score X to which it corresponds. Solving the z score formula for X we obtain

$$\overline{X} = z_\mu S_{\overline{X}} + \mu_0 \qquad (2.21)$$

or

$$\overline{X} = 1.645\,(2) + 50 = 53.29$$

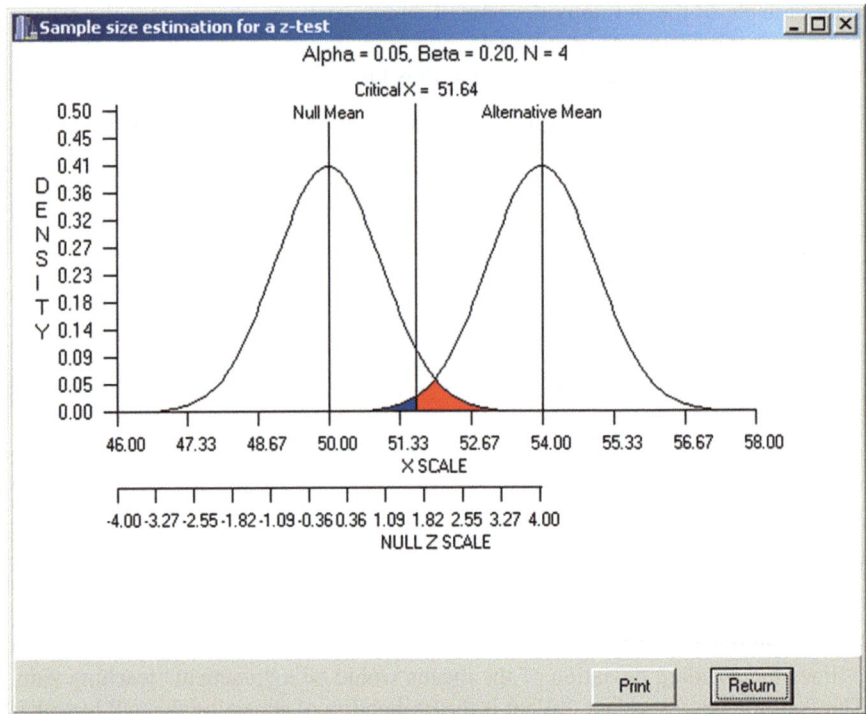

Fig. 2.2 Sample size estimation

Now that we have the raw score mean for the critical value of alpha, we can calculate the corresponding z score under the alternate distribution, that is

$$z_1 = \frac{\overline{X} - \mu_1}{S_{\overline{X}}} = \frac{53.29 - 54}{2} = -.355 \tag{2.22}$$

We may now ask, "What is the probability of obtaining a unit normal z score less than or equal to -0.355?" Using a table of the normal distribution or a program to obtain the cumulative probability of the z distribution we observe that the probability is $\beta = 0.359$. In other words, the probability of obtaining a z score of -0.355 or less is 0.359 under the normal distribution. We conclude then that the Type II error of our test, that is, the probability of incorrectly accepting the null hypothesis when, in fact, the true population mean is 54 is 0.359. Note that this nearly 36% chance of an error is considerably larger than the 5% chance of making the Type I error!

The sensitivity of our statistical test to detect true differences from the null hypothesized value is called the Power of our test. It is obtained simply as $1-\beta$. For the situation of detecting a difference as large as 4 (two standard deviations of the sample mean) in our previous example, the power of the test was

Determining Type II Error and Power of the Test

$1 - 0.359 = 0.641$. We may, of course, determine the power of the test for many other alternative hypotheses. For example, we may wish to know the power of our test to be sensitive to a discrepancy as large as 6 X score units of the mean. The figure below illustrates the power curves for different Type I error rates and differences from the null hypothesis.

Again, our procedure for obtaining the power would be

(a) Obtain the raw X-score mean corresponding to the critical value of α (region of rejection) under the null hypothesis. That is

$$\overline{X} = z_\alpha S_{\overline{X}} + \mu_0$$
$$= 1.645\,(2) + 50 = 53.29 \qquad (2.23)$$

(b) Obtain the z_1 score equivalent to the critical raw score for the alternate hypothesized distribution, e.g.

$$z_1 = (\overline{X} - \mu_1) / S_{\overline{X}}$$
$$= (53.29 - 56) / 2$$
$$= -2.71 / 2$$
$$= -1.355 \qquad (2.24)$$

(c) Determine the probability of obtaining a more extreme value than that obtained in (b) under the unit-normal distribution, e.g.

$$P\,(z<z_1|\ ND:\ \mu=0,\ \sigma=1)$$
$$= P\,(z<-1.355\ |\ ND:\ \mu=0,\ \sigma=1) = .0869 \qquad (2.25)$$

(d) Obtain the power as $1 - \beta = 1.0 - .0869 = .9131$ \qquad (2.26)

One may repeat the above procedure for any number of alternative hypotheses and plot the results in a figure such as that shown above. The above plot was made using the OpenStat option labeled "Generate Power Curves" in the Utilities menu.

As the critical difference increases, the power of the test to detect the difference increases. Minimum power is obtained when the critical difference is equal to zero. At that point power is equal to α, the Type I error rate. A different "power curve" may be constructed for every possible value of α. If larger values of α are selected, for example 0.20 instead of 0.05, then the test is more powerful for detecting true alternative distributions given the same meaningful effect size, standard deviation and sample size.

The Fig. 2.3 below shows the power curves for our example when selecting the following values of α: 0.01, 0.05, and 0.10.

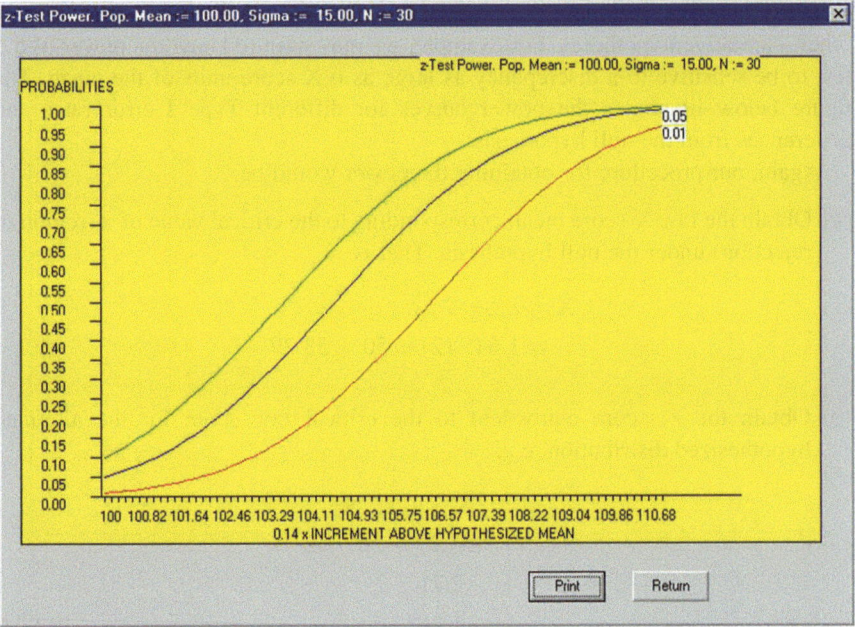

Fig. 2.3 Power curves

Sample Size Requirements for the Test of One Mean

The translation of a raw score mean into a standard score was obtained by

$$z = \frac{\overline{X} - \mu}{S_{\overline{X}}} \tag{2.27}$$

Likewise, the above formula may be rewritten for translating a z score into the raw score mean by:

$$\overline{X} = S_{\overline{X}} z + \mu \tag{2.28}$$

Now consider the distribution of an infinite number of sample means where each mean is based on the same number of randomly selected cases. Even if the original scores are not from a normally distributed population, if the means are obtained from reasonably large samples (N>30), the means will tend to be normally distributed. This phenomenon is known as the *Central Limit Theorem* and permits us to use the normal distribution model in testing a wide range of hypotheses concerning sample means.

Sample Size Requirements for the Test of One Mean

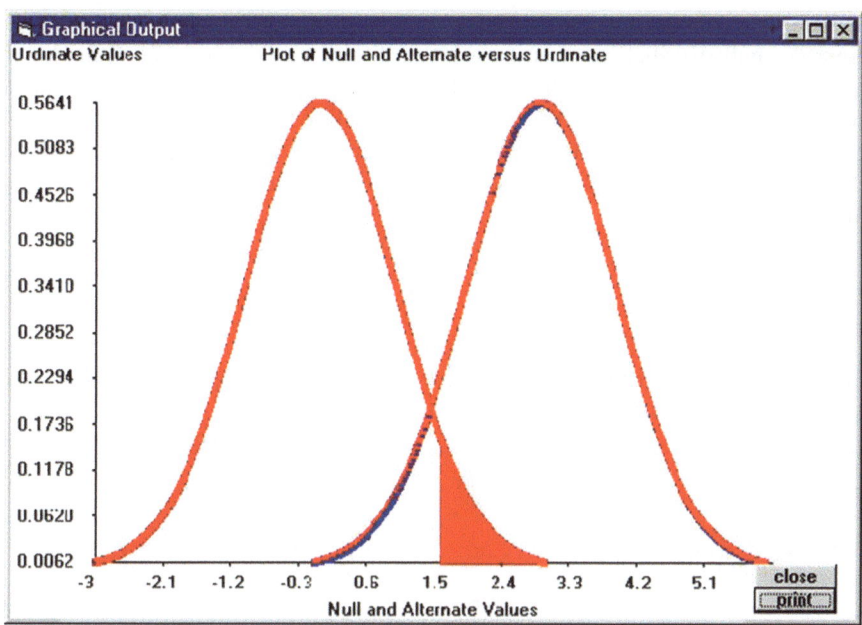

Fig. 2.4 Null and alternate hypotheses for sample means

The extreme "tails" of the distribution of sample means are sometimes referred to as "critical regions." Critical regions are defined as those areas of the distribution which are extreme, that is unlikely to occur often by chance, and which represent situations where you would reject the distribution as representing the true population should you obtain a sample in that region. The size of the region indicates the proportion of times sample values would result in rejection of the null hypothesis by chance alone—that is, result in a "Type I" error. For the situation of our last example, the full region "R" of say 0.05 may be split equally between both tails of the distribution, that is, 0.025 or R/2 is in each tail. For normally distributed statistics a 0.025 extreme region corresponds to a z score of either -1.96 for the lower tail or $+1.96$ for the upper tail. The critical sample mean values that correspond to these regions of rejection are therefore

$$\overline{X}_c = \pm \sigma_{\overline{X}} z_{\alpha/2} + \mu_0 \qquad (2.29)$$

In addition to the possibility of a critical score ($\overline{X}_c$) being obtained by chance part of the time (α) there also exists the probability (β) of accepting the null hypothesis when in fact the sample value is obtained from a population with a mean different from that hypothesized. Carefully examine the Fig. 2.4 above.

This figure represents two population distributions of means for a variable. The distribution on the left represents the null hypothesized distribution. The distribution on the right represents an alternate hypothesis, that is, the hypothesis that a sample mean obtained is representative of a population in which the mean differs from the null distribution mean be a given difference D. The area of this latter distribution to the left of the shaded alpha area of the left curve and designated as ß represents the chance occurrence of a sample falling within the region of acceptance of the null hypothesis, even when drawn from the alternate hypothesized distribution. The score value corresponding to the critical mean value for this alternate distribution is:

$$\overline{X}_c = \sigma_{\overline{X}} z_\beta + \mu_1 \tag{2.30}$$

Since formulas (2.29) and (2.30) presented above are both equal to the same critical value for the mean, they are equal to each other! Hence, we may solve for N, the sample size required in the following manner:

$$\sigma_{\overline{X}} z_\alpha + \mu_0 = \sigma_{\overline{X}} z_\beta + \mu_1 \tag{2.31}$$

where $\mu_1 = \mu_0 - D$

$$\text{and } \sigma_{\overline{X}} = \sigma_x / \sqrt{N} \tag{2.32}$$

Therefore,

$$(\sigma_x/\sqrt{N}) z_\alpha + \mu_0 = (\sigma_x/\sqrt{N}) z_\beta + \mu_1 \tag{2.33}$$

$$\text{or } \mu_1 - \mu_0 = (\sigma_x/\sqrt{N}) z_\alpha - (\sigma_x/\sqrt{N}) z_\beta \tag{2.34}$$

$$\text{or } D = \sigma_x/\sqrt{N} \, (z_\alpha - z_\beta) \tag{2.35}$$

$$\text{or } \sqrt{N} = (\sigma_x/D)(z_\alpha - z_\beta) \tag{2.36}$$

Note: z_β is a negative value in the above drawing because we are showing an alternative hypothesis above the null hypothesis. For an alternative hypothesis below the null, the result would yield an equivalent formula.

By squaring both sides of the above equation, we have an expression for the sample size N required to maintain both the selected α rate and ß rate of errors, that is

$$N = \frac{\sigma_x^2}{D^2} (z_\alpha + z_\beta)^2 \tag{2.37}$$

To demonstrate this formula (2.37) let us use the previous example of the teacher's experiment concerning a potentially superior teaching method. Assume that the teachers have agreed that it is important to contain both Type I error (α) and Type II error (ß) to the same value of 0.05. We may now determine the number of students that would be required to teach under the new teaching method and test.

Remember that we wished to be sensitive to a difference between the population mean of 50 by at least 4 points in the positive direction only, that is, we must obtain a mean of at least 54 to have a meaningful difference in the teaching method. Since this is a "one-tailed" test, α will be in only one tail of the null distribution. The z score which corresponds to this α value is 1.645. Similarly the value of z corresponding to the ß level of 0.05 is also 1.645. The sample size is therefore obtained as

$$N = \frac{10^2}{4^2}(1.645 + 1.645)^2$$
$$= (100/16)\,(3.29)^2 = (100/16) * 10.81 = 67.65$$

or approximately 68 students.

Clearly, to provide control over both Type I and Type II error, our research is going to require a larger sample size than originally anticipated! In this situation, the teacher could simply repeat the teaching with his new method with additional sections of students or accept a higher Type II error.

It is indeed a sad reflection on much of the published research in the social sciences that little concern has been expressed for controlling Type II error. Yet, as we have seen, Type II error can often lead to more devastating costs or consequences than the Type I error which is usually specified! Perhaps most of the studies are restricted to small available (non-random) samples, or worse, the researcher has not seriously considered the costs of the types of error. Clearly, one can control both types of error and there is little excuse for not doing so!

Confidence Intervals for a Sample Mean

When a mean is determined from a sample of scores, there is no way to know anything certain about the value of the mean of the population from which the sample was drawn. We do know however sample means tend to be normally distributed about the population mean. If an infinite number of samples of size n were drawn at random, the means of those samples would themselves have a mean μ and a standard deviation of σ/√n . This standard deviation of the sample means is called the standard error of the mean because it reflects how much in error a sample mean is in estimating the population mean μ on the average. Knowing how far sample means tend to deviate from μ in the long run permits us to state with some confidence what the likelihood (probability) is that some interval around our single sample mean would actually include the population mean μ.

Since sample means do tend to be normally distributed about the population mean, we can use the unit-normal z distribution to make confidence statements about our sample mean. For example, using the normal distribution tables or programs, we can observe that 95% of normally distributed z scores have values between −1.96 and + 1.96. Since sample means are assumed to be normally distributed, we may say that 95% of the sample means will surround the population mean μ in the interval of ± 1.96 the standard error of the means. In other words, if we draw a random sample of size n from a population of scores and calculate the sample mean, we can say with 95% confidence that the population mean is in the interval of our sample mean plus or minus 1.96 times the standard error of the means. Note however, that μ either is or is not in that interval. We cannot say for certain that μ is in the interval—only that we are some % confident that it is!

The calculation of the confidence interval for the mean is usually summarized in the following formula:

$$CI_\% = \overline{X} \pm z_\% \, \sigma_{\overline{X}} \tag{2.38}$$

Using our previous example of this chapter, we can calculate the confidence interval for the sample mean of 55 and the standard error for the sample of 25 subjects = 2 as

$$CI_{95} = \overline{X} \pm (1.96) \, 2$$
$$= 51.08 \text{ to } 58.92 \tag{2.39}$$

We state therefore that we are 95% confident that the actual population mean is between 58.92 and 51.08. Notice that the hypothesized mean (50) is not in this interval! This is consistent with our rejection of that null hypothesis. Had the mean of the null hypothesis been "captured" in our interval, we would have accepted the null hypothesis.

Another way of writing (2.39) above is

$$\text{probability } (\overline{X} - z_1 \sigma_{\overline{X}} < \mu < \overline{X} + z_2 \sigma_{\overline{X}}) = P \tag{2.40}$$

where z_1 and z_2 are the z scores corresponding to the lower and upper values of the % confidence desired, and P is the probability corresponding to the % confidence.

For example we might have written our results of the teacher experiment as

$$\text{probability } [(55 - 1.96(2) < \mu < 55 + 1.96(2)] = .95$$
$$\text{or } \text{probability } (51.08 < \mu < 58.92) = .95$$

Frequency Distributions

A variable is some measure or observation of an attribute that varies from subject to subject. We are frequently interested in the shape of the distribution of the frequencies of objects who's scores fall in each category or interval of our variable. When the shape of the frequency distribution closely resembles that of a theoretical

Frequency Distributions

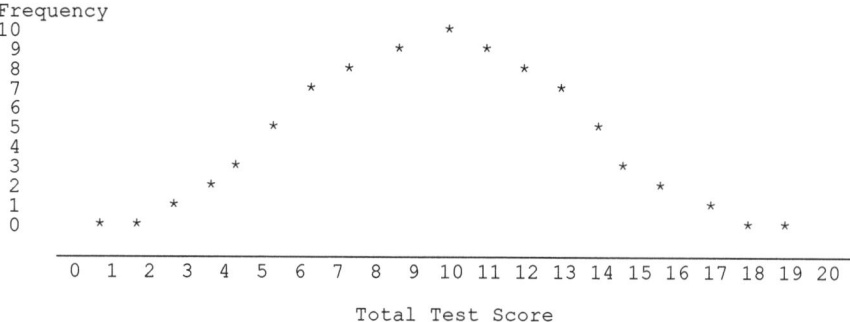

Fig. 2.5 Sample plot of test scores

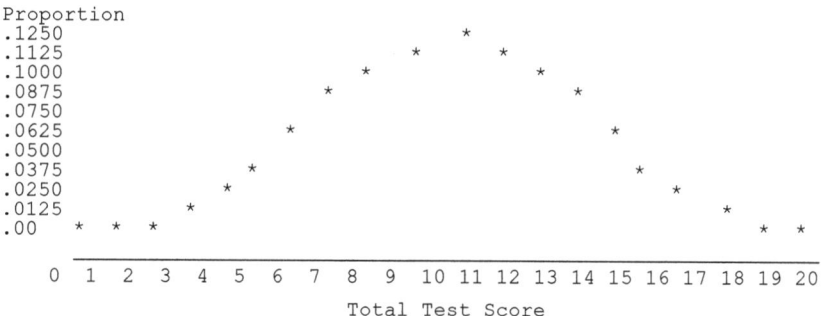

Fig. 2.6 Sample proportions of test scores

model of such distributions, we may utilize statistics developed for those theoretical distributions to describe our observations. We will examine some of the most common theoretical distributions. First, let us consider a simple figure representing the frequency of scores found in intervals of a classroom teacher's test. We will assume the teacher has administered a 20 item test to 80 students and has "plotted" the number of students obtaining the various total scores possible. The plot might look as above (Fig. 2.5):

We can also express the number of subjects in each score range as a proportion of the total number of observations. For example, we could divide each of the frequencies above by 80 (the number of observations) and obtain (Fig. 2.6):

If the above distribution of the proportion of test scores at each possible value had been obtained on a very, very large number of cases in a population of subjects, we would refer to the proportions as probabilities. We would then be able to make statements such as "the probability of a student earning a score of 10 in the population is 0.01."

Sometimes we draw a figure that represents the cumulative frequencies divided by the total number of observations. For example, if we accumulate the frequencies

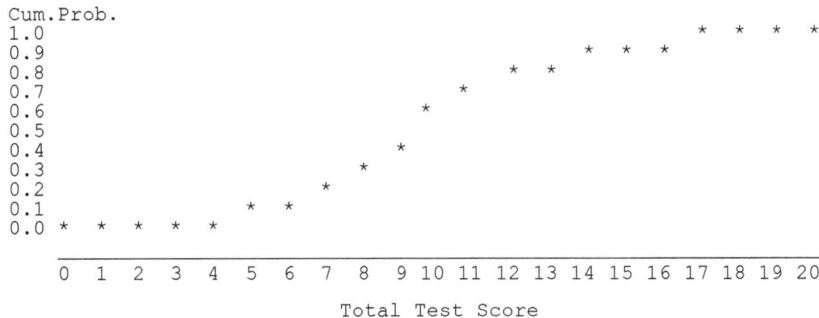

Fig. 2.7 Sample sumulative probabilities of test scores

represented in the previous figure the cumulative distribution would appear as (Fig. 2.7):

If the above 80 observations constituted the population of all possible observations on the 20 item test, we have no need of statistics to estimate population parameters. We would simply describe the mean and variance of the population values. If, on the other hand, the above 80 scores represents a random sample from a very, very large population of observations, we could anticipate that another sample of 80 cases might have a slightly different distribution appearance. The question may now be raised, what is a reasonable "model" for the distribution of the population of observations? There are clearly a multitude of distribution shapes for which the above sample of 80 scores might be reasonably thought to be a sample. Because we do not wish to examine all possible shapes that could be considered, we usually ask whether the sample distribution could be reasonably expected to have come from one of several "standard" distribution models. The one model having the widest application in statistics is called the "Normal Distribution". It is that model which we now examine.

The Normal Distribution Model

The Normal Distribution model is based on a mathematical function between the height of a probability curve for each possible value on the horizontal axis. Since the horizontal axis reflects measurement values, we must first translate our observations into "standard" units that may be used with any set of observations. The "z" score transformation is the one used, that is, we standardize our scores by dividing a scores deviation from the mean by the standard deviation of the scores. If we know the population mean and standard deviation, the transformation is

$$z_i = \frac{(X_i - \overline{X})}{\sigma_x} \quad (2.41)$$

If the population mean and standard deviation are unknown, then the sample estimates are used instead.

The Normal Distribution function (also sometimes called the Gaussian distribution function) is given by

$$h = \frac{1}{\sqrt{2\pi}} e^{\frac{-z^2}{2}} \tag{2.42}$$

where h is the height of the curve at the value z and e is the constant 2.7182818.....

To see the shape of the normal distribution for a large number of z scores, select the Analysis option and move the cursor to the Miscellaneous option. A second menu will appear. Click on the Normal Distribution Curve option. Values of h are drawn for values between approximately -3.0 to $+3.0$. It should be noted that the normal distribution actually includes values from -infinity to + infinity. The area under the normal curve totals 1.0. The area between any two z scores on the normal distribution therefore reflect the proportion (or probability) of z scores in that range. Since the z scores may be ANY value from -infinity to + infinity, the normal distribution reflects observations made on a scale considered to yield continuous scores.

THE TRUE BELL CURVE - The distribution of SUCCESS in life in relationship to AGE follows a true bell curve:
  At age 5 success is not peeing in your pants
  At age 10 success is having friends
  At age 16 success is having your driver's license
  At age 20 success is having sex
  At age 35 success is having money
  At age 50 success is having money
  At age 65 success is having sex
  At age 70 success is having your driver's license
  At age 75 success is having friends
  At age 80 success is not peeing in your pants

The Median

While the mean is obtained as the average of scores in a distribution, it is not the only measure of "central tendency" or statistic descriptive of the "typical" score in a distribution. The *median* is also useful. It is the "middle score" or that value below which lies 50% of the remaining score values. The difference between the mean and median values is an indicator of how "skewed" are the distribution of scores. If the difference is positive (mean greater than the median) this would indicate that the mean is highly influenced by "extremely" high scores. If you plot the distribution of scores, there is typically a "tail" extending to the right (assuming the scores are arranged with low scores to the left and higher scores to the right.) We would say the distribution is positively skewed. When the distribution is negatively skewed the mean is less than the median. The median is highly useful for describing the typical score when the distribution is highly skewed. For example, the average income in the United States is much greater than the median income. A few millionaires (or billionaires) in the population skews the distribution. In this case, the median is more "representative" of the "typical" income.

Skew

The *skew* of a distribution is obtained as the third moment of the distribution. The first moment, the mean, is the average of the scores (sum of X's divided by the number of X's.) The second moment is the variance and is the average of the squared deviations from the mean. The third moment is the average of the cubed deviations from the mean. We can write this as:

$$\text{Skew} = \frac{\sum (X - \mu)^3}{N} \tag{2.43}$$

Professor: "OK students, you have fifteen minutes to plot the bivariate distribution between A and B, fifteen minutes to compute the correlation between A and B, and 5 SECONDS to compute the kurtosis of B." One student stands up very worried: "Excuse me Professor, how can we posssibly compute a kurtosis in 5 SECONDS?" The Professor looks at the class very reassuring: "No need to be worried, kids, IT TAKES ONLY A MOMENT!!"

Kurtosis

What do statistics professors get when they drink too much? Kurtosis of the Liver!

A distribution may not only be skewed (not bell-shaped) but may also be "flatter" or more "peaked" than found in the normal curve. When a distribution is more flat we say that it is *platykurtik*. When it is more peaked we say it is *leptokurtik*. When it follows the typical normal curve it is described as *mesokurtik*. Kurtosis therefore describes the general height of the distribution across the score range. The kurtosis is obtained as the fourth moment about the mean. We can write it as:

$$\text{Kurtosis} = \frac{\sum (X - \mu)^4}{N} \tag{2.44}$$

A middle aged man suddenly contracted the dreaded disease kurtosis. Not only was this disease severely debilitating but he had the most virulent strain called leptokurtosis.

A close friend told him his only hope was to see a statistical physician who specialized in this type of disease. The man was very fortunate to locate a specialist but he had to travel 800 miles for an appointment. After a thorough physical exam, the statistical physician exclaimed, "Sir, you are indeed a lucky person in that the FDA has just approved a new drug called mesokurtimide for your illness. This drug will bulk you up the middle, smooth out your stubby tail, and restore your longer range of functioning. In other words, you will feel "NORMAL" again!"

The Binomial Distribution

Some observations yield a simple dichotomy that may be coded as 0 or 1. For example, you may draw a sample of subjects and observe the gender of each subject. A code of 1 may be used for males and 0 for females (or vice-versa). In a population of such scores, the proportion of observations coded 1 (P) is the mean (θ) of the scores. The population variance of dichotomous scores is simple $\theta(1-\theta)$ or $P(1-P)$. When a sample is drawn from a population of dichotomous scores, the sample mean, usually denoted simply as p, is an estimator of θ and the population variance is estimated by $p(1-p)$. The probability of observing a specific number of subjects that would be coded 1 when sampling from a population in which the proportion of such subjects is P can be obtained from

$$X = \frac{N!}{n!(N-n)!} p^n (1-p)^{N-n}$$

or simply

$$X = \binom{N}{n} p^n (1-p)^{N-n} \quad (2.45)$$

where X is the probability,
N is the size of the sample,
n is the number of subjects coded 1 and
P is the population proportion of objects coded 1.

The ! symbol in the above equation is the "factorial" operation, that is, n! means $(1)(2)(3)....(n)$, the product of all integers up through n. Zero factorial is defined to be equal to 1, that is, $0! = 1$.

For any sample of size N, we can calculate the probabilities of obtaining 0, 1, 2, ... , n values of the objects coded 1 when the population value is P. Once those values are obtained, we may also obtain the cumulative probability distribution. For example, assume you are sampling males and females from a population with a mean of 0.5, that is, the number of males (coded 1) equals the number of females

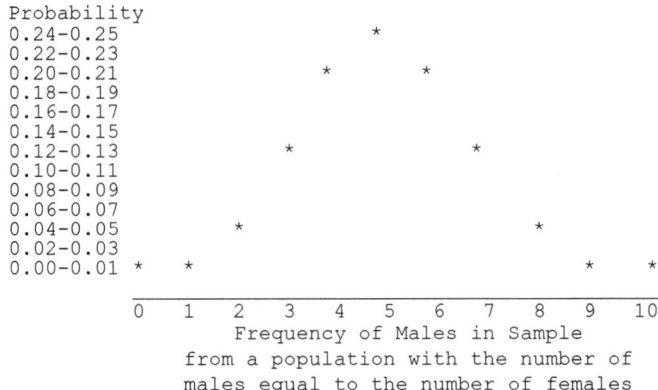

Fig. 2.8 Sample probability plot

(coded 0). Now assume you randomly select a sample of 10 subjects and count the number of males (n). The probabilities for n = 0, 1, ..., N are as follows:

No. Males Observed	Probability	Cumulative Probability
0	0.00097	0.00097
1	0.00977	0.01074
2	0.04395	0.05469
3	0.11719	0.17188
4	0.20508	0.37695
5	0.24609	0.62405
6	0.20508	0.82813
7	0.11719	0.94531
8	0.04395	0.98926
9	0.00977	0.99902
10	0.00097	1.00000

Now let us plot the above binomial distribution (Fig. 2.8):

A man who travels a lot was concerned about the possibility of a bomb on board his plane. He determined the probability of this, found it to be low but not low enough for him. So now he always travels with a bomb in his suitcase. He reasons that the probability of two bombs being on board would be infinitesimal.

The Poisson Distribution

The Poisson distribution describes the frequency with which discrete binomial events occur. For example, each child in a school system is either in attendance or not in attendance. The probability of each child being absent is, however, quite small. The probability of X children being absent from a school increases with the size of the school (n). Another example is in the area of school drop-outs. Each student may be considered to be either a drop-out or not a drop-out. The probability of being a drop-out student is usually quite small. The probability that X students

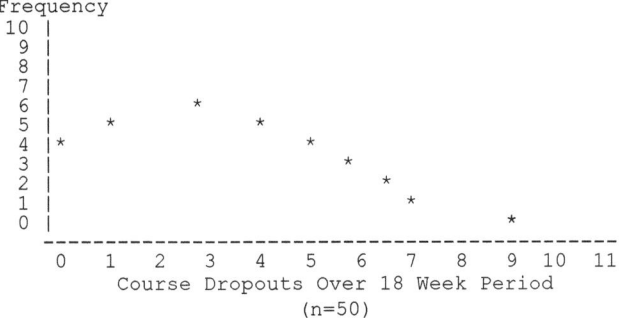

Fig. 2.9 A poisson distribution

out of n drop out over a given period of time may also be described by the Poisson distribution.

The figure above illustrates a representative Poisson distribution (Fig. 2.9):

The frequency (height) of the Poisson distribution is obtained from the following function:

$$f(x) = \frac{L^x e^{-L}}{x!} \qquad (2.46)$$

where $\mu = L$, the mean of the population distribution and $\sigma = L =$ the standard deviation of the population distribution

We note that when a variable (e.g. dropouts occurring) has a mean and standard deviation that are equal in the sample, the distribution may fit the Poisson model. In addition, it is important to remember that the variable (X) is a discrete variable, that is, only consists of integer values.

The Chi-Squared Distribution

In the field of statistics there is another important distribution that finds frequent use. The chi-squared statistic is most simply defined as the square of a normally distributed z score. Referring back to the paragraph on z scores, you will remember that is obtained as

$$z_i = \frac{(X_i - \overline{X})}{\sigma_Y} \qquad (2.47)$$

that is, the deviation from the mean divided by the variance in the population of normally distributed scores. The z scores in an infinite population of scores ranges from -4 to + 4 . If we square randomly selected z scores, all resulting values are

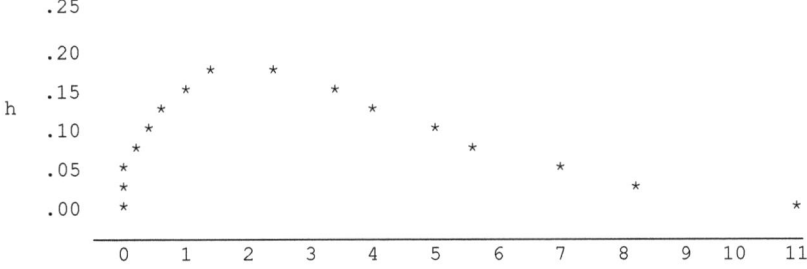

Fig. 2.10 Chi-squared distribution with 4° of freedom

greater than or equal to zero. If we randomly select n z-scores, squaring each one, the sum of those squared z scores is defined as a Chi-Squared statistic with n degrees of freedom. Each time we draw a random sample of n z-scores and calculate the Chi-squared statistic, the value may vary from sample to sample. The distribution of these sample Chi-squared statistics follows the distribution density (height) function:

$$h = \frac{\chi^{\frac{n-2}{2}} e^{\frac{-\chi}{2}}}{\left[2^{\frac{n}{2}} \Gamma\left(\frac{n}{2}\right)\right]^{-1}} \tag{2.48}$$

where χ is the Chi-squared statistic,
n is the degrees of freedom,
e is the constant 2.7181... of the natural logarithm,
and $\Gamma()$ is the gamma function.

In the calculation of the height of the chi-squared distribution, we encounter the gamma function (Γ). The gamma function is similar to another function, the factorial function (n!). The factorial of a number like 5, for example, is $5 \times 4 \times 3 \times 2 \times 1$ which equals 80. The factorial however only applies to integer values. The gamma function however applies to continuous values as well as integer values. You can approximate the gamma function however by interpolating between integer values of the factorial. For example, the value of $\Gamma(4)$ is equal to 3! or $3 \times 2 \times 1 = 6$. In general, $\Gamma(k-1) = k!$.

A sample distribution of Chi-squared statistics with 4° of freedom is illustrated above (Fig. 2.10)

The F Ratio Distribution

Another sample statistic which finds great use in the field of statistics is the F statistic. The F statistic may be defined in terms of the previously defined Chi-squared statistic. It is the ratio of two independent chi-squares, each of which has been divided by its degrees of freedom, that is

$$F_{(n_1,n_2)} = \frac{\frac{\chi^2}{n_1}}{\frac{\chi^2}{n_2}} \quad (2.49)$$

where χ^2 is the chi-squared statistic, and n_1 and n_2 are the degrees of freedom for the numerator and denominator chi-squares.

As before, we can develop the theoretical model for the sampling distribution of the F statistic. That is, we assume we repeatedly draw independent samples of n1 and n2 normally distributed z-scores, square each one, sum them up in each sample, and form a ratio of the two resulting chi-squared statistics. The height (density) function is given as

$$h = \frac{\Gamma\left[\frac{n_1+n_2}{2}\right] n_1^{\frac{n_1}{2}} n_2^{\frac{n_2}{2}}}{\Gamma\left(\frac{n_1}{2}\right)\Gamma\left(\frac{n_2}{2}\right)} \cdot \frac{F^{\frac{n_1-2}{2}}}{[n_1 F + n_2]^{\frac{n_1+n_2}{2}}} \quad (2.50)$$

where F is the sample statistic,
n_1 and n_2 are the degrees of freedom, and
$\Gamma()$ is the gamma function described in the previous paragraph.

An example of the distribution of the F statistic for n_1 and n_2 degrees of freedom may be generated using the Distribution Plots and Critical Value procedure from the Simulation menu in your OpenStat package.

The "Student" t Test

The z statistic used to test hypotheses concerning sample means assumes the use of the normal distribution. However we have seen that the unbiased estimate of the standard deviation of the sample means is "adjusted" for the sample size, that is $S/\sqrt{(N-1)}$. If N is large, the distribution we can normally assume the distribution of the means is normal. When N is small, the "fit" to the normal distribution is less likely. William Gosset (who published under the name "Student") developed the mathematics for distributions that differ for the size of N but approach the normal (Gaussian) distribution as N increases in size. We obtain our statistic t in the same manner that we did for the z tests but instead of using the normal distribution, we use the t distribution. This distribution is described by the following equation:

$$y = \frac{C}{[1 + (t^2/df)]^{(df+1)/2}} \quad (2.51)$$

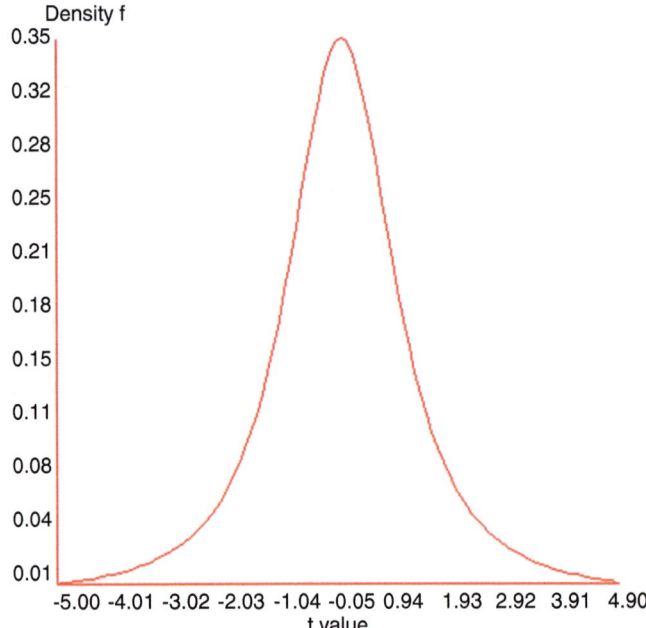

Fig. 2.11 t distribution with 2° of freedom

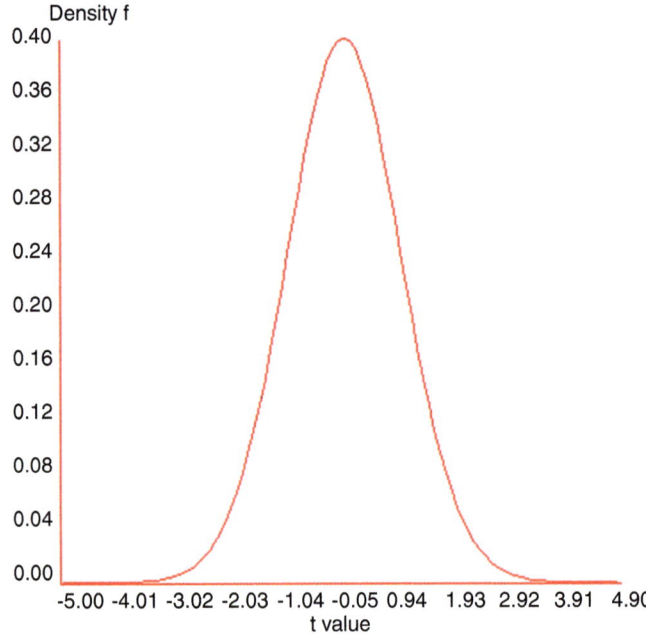

Fig. 2.12 t distribution with 100° of freedom

The "Student" t Test

Where

$$C = \frac{[(df-1)/2]!}{\sqrt{(ndf)}[(df-2)/2]!}$$

Note: df is a single value for "degrees of freedom"

Shown above are two t distribution plots, the first with 2° of freedom and the second with 100° of freedom (Figs. 2.11 and 2.12):

If you examine the density (height) of the curve on each of the above plots, you will see that the density is much greater for the plot with only 2 degrees of freedom. The "tails" of the t distribution are greater as the degrees of freedom decrease. If one is testing a hypothesis at the alpha level of say 0.05, it will take a larger value of t in a t-test in comparison to a z test to be significant for the smaller samples! The degrees of freedom for the t-test will vary depending on the nature of the hypothesis being tested.

Chapter 3
The Product Moment Correlation

It seems most living creatures observe relationships, perhaps as a survival instinct. We observe signs that the weather is changing and prepare ourselves for the winter season. We observe that when seat belts are worn in cars that the number of fatalities in car accidents decrease. We observe that students that do well in one subject tend to perform will in other subjects. This chapter explores the linear relationship between observed phenomena.

If we make systematic observations of several phenomena using some scales of measurement to record our observations, we can sometimes see the relationship between them by "plotting" the measurements for each pair of measures of the observations. As a hypothetical example, assume you are a commercial artist and produce sketches for advertisement campaigns. The time given to produce each sketch varies widely depending on deadlines established by your employer. Each sketch you produce is ranked by five marketing executives and an average ranking produced (rank 1 = best, rank 5 = poorest.) You suspect there is a relationship between time given (in minutes) and the average quality ranking obtained. You decide to collect some data and observe the following (Fig. 3.1):

Average rank (Y)	Minutes (X)
3.8	10
2.6	35
4.0	5
1.8	42
3.0	30
2.6	32
2.8	31
3.2	26
3.6	11
2.8	33

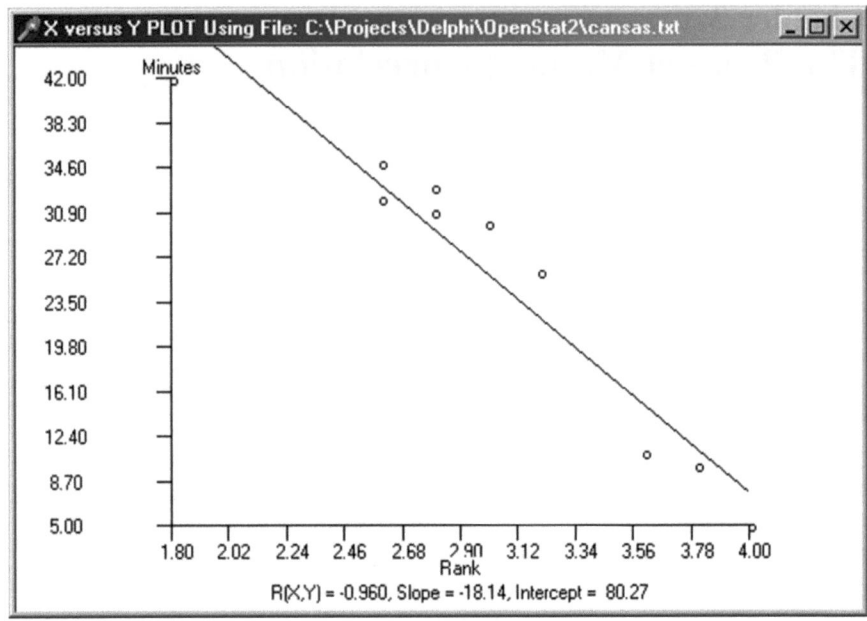

Fig. 3.1 A negative correlation plot

Testing Hypotheses for Relationships Among Variables: Correlation

Scattergrams

While the mean and standard deviation of the previous chapter are useful for describing the central tendency and variability of the measures of a single variable, there are frequent situations in which it is desirable to obtain an index of how values measured on TWO variables tend to vary in the same or opposite directions. This "co-variability" of two variables may be visually represented by means of a Scattergram, for example, the figure below represents a scattergram of individual's scores on two variables, X and Y (Fig. 3.2).

In the above figure, each asterisk (*) represents a subject's position on two scales of measurement - on the X scale and on the Y scale. We can observe that subjects with larger X score values *tend* to have larger Y score values.

Now consider a set of score pairs representing measurements on two variables, College Grade Point Average (GPA) and Perceptions of Inadequacy (PI). The figure below the data represents the scattergram of subject scores (Fig. 3.3).

Testing Hypotheses for Relationships Among Variables: Correlation

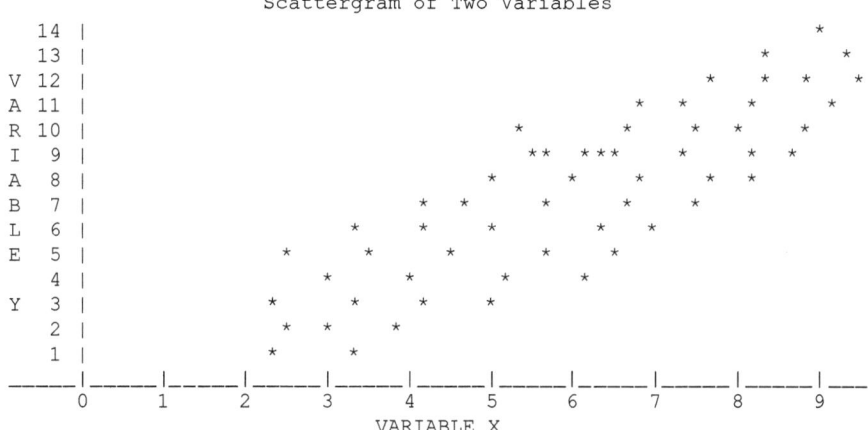

Fig. 3.2 Scattergram of two variables

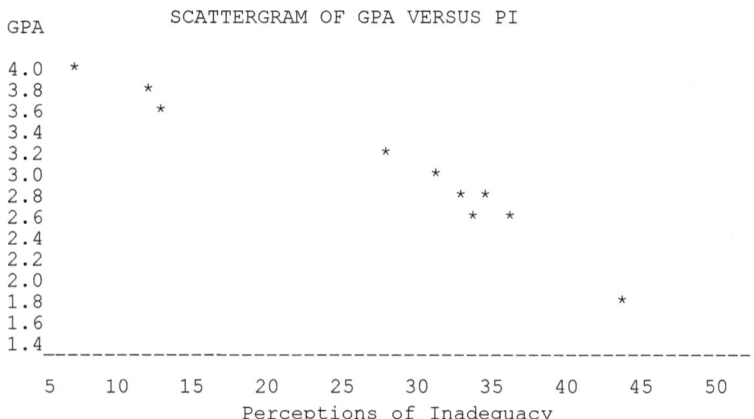

Fig. 3.3 Scattergram of a negative relationship

Subject	GPA	PI
1	3.8	10
2	2.6	35
3	4.0	5
4	1.8	42
5	3.0	30
6	2.6	32
7	2.8	31
8	3.2	26
9	3.6	11
10	2.8	33

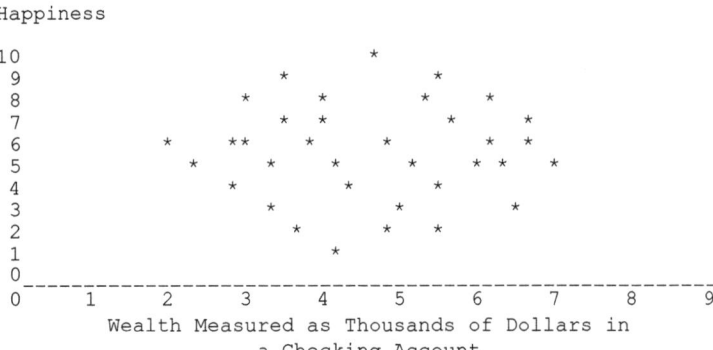

Fig. 3.4 Scattergram of two variables with low relationship

In this example there is a negative relationship between the two variables, that is, as a subject's perceptions of inadequacy increase, there is a corresponding decrease in grade point average! (The data are hypothetical if you hadn't guessed).

Many variables, of course, may not be related at all. In the below scattergram, there is no systematic co-variation between the two variables (Fig. 3.4):

A simple way to construct an index of the relationship between two variables might be to simply average the product of the score pairs for the individuals. Unfortunately, the size of this index would vary as a function of the size of the numbers yielded by our measurement scales. We wouldn't be able to compare the index we obtained for, say, grade point averages in high school and college with the index we would obtain for college grade point averages and beginning salaries! On the other hand, an average of score products might be useful if we first transformed all of our measurements to a COMMON scale of measurement. In fact, this is just what Pearson did! By converting scores to a scale of measurement such that the mean score is always zero and the standard deviation of the scores on a variable is always 1.0, he was able to produce an index which, for any pair of variables, always varies between -1.0 and $+1.0$!

Transformation to z Scores

We define a z score as a simple linear transformation of raw scores which involves the formula

$$z_i = \frac{(X_i - \bar{X})}{S_x} \qquad (3.1)$$

where z_i is the z score for an individual, X_i the individual's raw score and S_x is the standard deviation of the set of X scores.

When we have a pair of scores for each individual, we must adopt some method for differentiating between the two variables. Often we simply name the variables X

Transformation to z Scores

and Y or X_1 and X_2. For the case of simple correlation discussed in this section, we will adopt the first method, i.e., the use of X and Y. We will use the second method when we start to deal with three or more variables at the same time.

The Pearson Product–moment correlation is defined as

$$r_{x,y} = \frac{\sum_{i=1}^{N} z_{x_i} z_{y_i}}{N} \tag{3.2}$$

that is, the average of z score products for the N objects or subjects in our sample. Note, we have used the BIASED standard deviation in our z score transformations (divided by N, not N−1).

Now let us see how we apply the above formula in obtaining a coefficient of correlation for the above scattergram. First, we must transform our raw scores (Y and X) to z scores. To do this we must obtain the mean and standard deviation for each variable. In the figure below we have obtained the mean and standard deviation of each variable, obtained the deviation of each score from the respective mean, and finally, divided each deviation score by the corresponding standard deviation. We have also shown the product of the z scores for both the X and Y variables. It is this product of z scores which, when averaged, yields the *product–moment correlation* coefficient!

case	Y	X	$(Y_i - \bar{Y})$	$(X_i - \bar{X})$	y^{z_i}	x^{z_i}	$y^{z_i} x^{z_i}$
1	3.8	10	.78	-15.5	1.253	-1.318	-1.651
2	2.6	35	-.42	9.5	-.675	.808	-.545
3	4.0	5	.98	-20.5	1.574	-1.743	-2.743
4	1.8	42	-1.22	16.5	-1.960	1.403	-2.750
5	3.0	30	.02	4.5	.032	.383	.012
6	2.6	32	-.42	6.5	-.675	.553	-.373
7	2.8	31	-.22	5.5	-.353	.468	-.165
8	3.2	26	.18	0.5	.289	.043	.012
9	3.6	11	.58	-14.5	.932	-1.233	-1.149
10	2.8	33	-.22	7.5	-.353	.638	-.225

$$\sum_{i=1}^{N} Y_i = 30.2 \qquad \bar{Y} = 3.02$$

$$\sum_{i=1}^{N} Y_i^2 = 95.08 \qquad S_y^2 = \sum_{i=1}^{N} Y^2 / N - \bar{Y}^2$$

$$= 9.508 - 9.1204 = 0.3876$$

$$\text{and } S_y = 0.62257529$$

$$\sum_{i=1}^{N} X_i = 255.0 \qquad \bar{X} = 25.5$$

$$\sum_{i=1}^{N} X_i^2 = 7885.0 \qquad S_x^2 = \sum_{i=1}^{N} X^2 / N - \bar{X}^2$$

$$= 788.5 - 650.25 = 138.25$$

$$\text{and } S_x = 11.757976$$

$$r_{yx} = \sum y^{z_i} x^{z_i} / N = -9.577 / 10 = -.958$$

The above method for obtaining the product–moment correlation is quite laborious and it is easy to make arithmetic mistakes and rounding errors. Let's look for another way which does not require actually computing the z scores for each variable. First, let us substitute the definition of the z scores in the formula for the correlation:

$$r_{x,y} = \frac{\sum_{i=1}^{N} z_{x_i} z_{y_i}}{N} = \frac{\sum_{i=1}^{N} \left(\frac{Y_i - \overline{Y}}{S_y}\right)\left(\frac{X_i - \overline{X}}{S_x}\right)}{N} = \frac{\sum_{i=1}^{N} \left[X_i Y_i - Y_i \overline{X} - \overline{Y} X_i + \overline{YX}\right]}{N S_y S_x} \quad (3.3)$$

or

$$r_{x,y} = \frac{\sum_{i=1}^{N} Y_i X_i - \sum_{i=1}^{N} Y_i \overline{X} - \sum_{i=1}^{N} \overline{Y} X_i + \sum_{i=1}^{N} \overline{YX}}{N S_y S_x}$$

$$= \frac{\sum_{i=1}^{N} Y_i X_i - \overline{X} \sum_{i=1}^{N} Y_i - \overline{Y} \sum_{i=1}^{N} X_i + N \overline{YX}}{N S_y S_x} \quad (3.4)$$

or

$$r_{x,y} = \frac{\frac{\sum_{i=1}^{N} Y_i X_i}{N} - \overline{X} \frac{\sum_{i=1}^{N} Y_i}{N} - \overline{Y} \frac{\sum_{i=1}^{N} X_i}{N} + \overline{YX}}{S_y S_x} = \frac{\frac{\sum_{i=1}^{N} Y_i X_i}{N} - \overline{XY} - \overline{YX} + \overline{YX}}{S_x S_y}$$

$$= \frac{\frac{\sum_{i=1}^{N} Y_i X_i}{N} - \overline{XY}}{S_x S_y} \quad (3.5)$$

The last formula does not require us to use z scores at all. We only need to use raw X and Y scores! Since we have already learned to compute S_x and S_y in terms of raw scores, we can do a little more algebra manipulation of the above formula and obtain

$$r_{x,y} = \frac{N \sum_{i=1}^{N} X_i Y_i - \left(\sum_{i=1}^{N} Y_i\right)\left(\sum_{i=1}^{N} X_i\right)}{\sqrt{\left[N \sum_{i=1}^{N} Y_i^2 - \left(\sum_{i=1}^{N} Y_i\right)^2\right]\left[N \sum_{i=1}^{N} X_i^2 - \left(\sum_{i=1}^{N} X_i\right)^2\right]}} \quad (3.6)$$

This formula is particularly advantages in that it utilizes the sums and sums of squared scores and the sum of cross-products of the X and Y scores. In addition, it contains fewer divisions which reduces round-off error! Using the previous example, we would obtain:

case	Y	X	Y²	X²	YX
1	3.8	10	14.44	100	38.0
2	2.6	35	6.76	1225	91.0
3	4.0	5	16.00	25	20.0
4	1.8	42	3.24	1764	75.6
5	3.0	30	9.00	900	90.0
6	2.6	32	6.76	1024	83.2
7	2.8	31	7.84	961	86.8
8	3.2	26	10.24	676	83.2
9	3.6	11	12.96	121	39.6
10	2.8	33	7.84	1089	92.4
	30.2	255	95.08	7885	699.8

$$r_{x,y} = \frac{(10)(699.8) - (30.2)(255)}{\sqrt{\left[10(95.08) - (30.2)^2\right]\left[10(7885) - (255)^2\right]}}$$

$$= \frac{6998 - 7701}{\sqrt{[950.8 - 912.04][78850 - 65025]}}$$

or

$$r_{x,y} = \frac{-703}{\sqrt{(38.76)(13825)}} = \frac{-703}{\sqrt{535857}} = \frac{-703}{732.02254}$$

$$= -0.960 \text{ (approximately)} \qquad (3.7)$$

Notice that the product–moment correlation obtained by this method differs by approximately .002 obtained in the average of z score products method. The first method had much more round-off error due to our calculations only being carried out to the nearest thousandths. Our results by this second method are clearly more accurate, even for only ten cases!

If you use the unbiased estimates of variances, other formulas may be written to obtain the product–moment correlation coefficient. Remember we divide the sum of squared deviations about the mean by N-1 for the unbiased estimate of population variance. In this case the average of z-score products is also divided by N-1 and by substituting the definition of a z score for both X and Y we obtain:

$$r_{x,y} = \frac{C_{x,y}}{s_x s_y} \qquad (3.8)$$

where

$$C_{x,y} = \frac{\sum_{i=1}^{N} X_i Y_i - \frac{\sum_{i=1}^{N} X_i \sum_{i=1}^{N} Y_i}{N}}{N-1}, \qquad (3.9)$$

the covariance of x and y and the unbiased estimates of variance are:

$$s_x^2 = \frac{\sum_{i=1}^{N} X_i^2 - \left(\sum_{i=1}^{N} X_i\right)^2 / N}{N-1} \qquad (3.10)$$

$$s_y^2 = \frac{\sum_{i=1}^{N} Y_i^2 - \left(\sum_{i=1}^{N} Y_i\right)^2 / N}{N-1} \qquad (3.11)$$

with

$$s_x = \sqrt{s^2_x} \text{ and } s_y = \sqrt{s^2_y}$$

To further understand and learn to interpret the product–moment correlation, OpenStat provides a means of simulating pairs of data, plotting those pairs, drawing the "best-fitting line" to the data points and showing the marginal distributions of the X and Y variables. Go to the Simulation menu and click on the Bivariate Scatter Plot. The figure below shows a simulation for a population correlation of −0.95 with population means and variances as shown. A sample of 100 cases are generated. Actual sample means and standard deviations will vary (as sample statistics do!) from the population values specified (Fig. 3.5).

```
POPULATION PARAMETERS FOR THE SIMULATION
Mean X := 100.000, Std. Dev. X := 15.000
Mean Y := 100.000, Std. Dev. Y := 15.000
Product-Moment Correlation := -0.900
Regression line slope := -0.900, constant := 190.000
SAMPLE STATISTICS FOR 100 OBSERVATIONS FROM THE POPULATION
Mean X :=  99.988, Std. Dev. X := 14.309
Mean Y := 100.357, Std. Dev. Y := 14.581
Product-Moment Correlation := -0.915
Regression line slope := -0.932, constant := 193.577
```

Simple Linear Regression

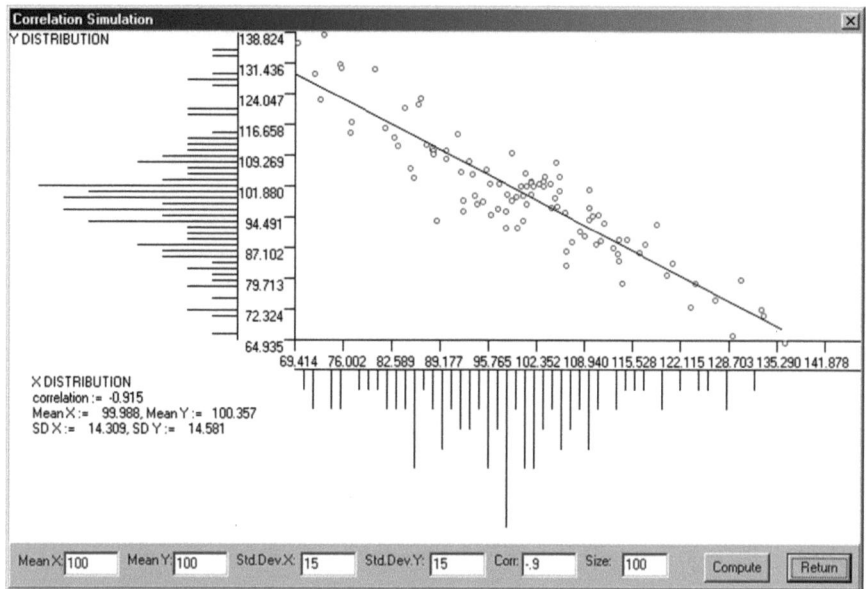

Fig. 3.5 A simulated negative correlation plot

Simple Linear Regression

The product–moment correlation discussed in the previous section is an index of the linear relationship between two continuous variables. But what is the nature of that linear relationship? That is, what is the slope of the line and where does the line intercept the vertical (Y variable) axis? This unit will examine the straight line "fit" to data points representing observations with two variables. We will also examine how this straight line may be used for prediction purposes as well as describing the relationship to the product–moment correlation coefficient.

To introduce the "straight line fit" we will first introduce the concept of "least-squares fit" of a line to a set of data points. To do this we will keep the number of X and Y score pairs small. Examine the figure below. It represents a set of five score pairs similar to those presented in the previous unit (Fig. 3.6).

In the figure, each point represents the intersection of X and Y score values for an observed case. Also shown is a line that represents the "best fitting line" to the data points:

```
          Case     1   2   3   4   5
          ------------------------
          X  |  1   2   3   4   5
          Y  |  2   1   3   5   4
          ------------------------
```

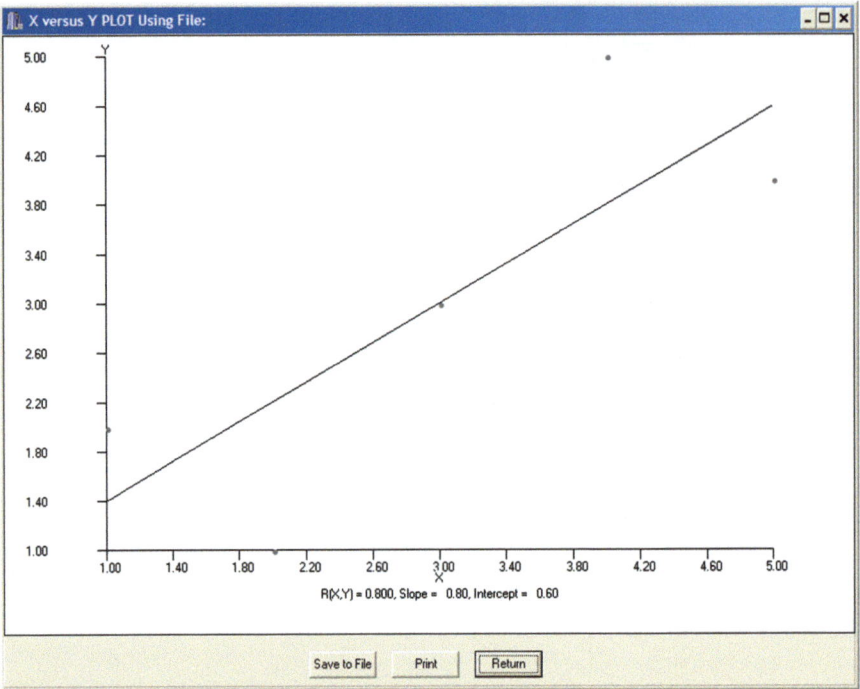

Fig. 3.6 X versus Y plot of five values

The Least-Squares Fit Criterion

In regression analysis, we want to develop a formula for a straight line which optimally predicts each Y score from a given X score. For example, if Y is a student's College Grade Point Average (GPA) and X is the high school grade point average (HSGPA), we wish to develop an equation which will predict the GPA given the HSGPA. Straight line formulas generally are of the form

$$Y = BX + C \qquad (3.12)$$

where B is the slope of the line, and C is a constant representing the point where the line crosses the Y axis. This is also called the intercept.

In the Figure below, B is the slope of the line (the number of Y units (rise) over 1 unit of X (run)). C is the intercept where the line crosses the Y axis (Fig. 3.7).

If X and Y scores are transformed to z scores using the transformations

$$z_x = (X_i - \bar{X})/\sigma_x \qquad (3.13)$$

The Least-Squares Fit Criterion

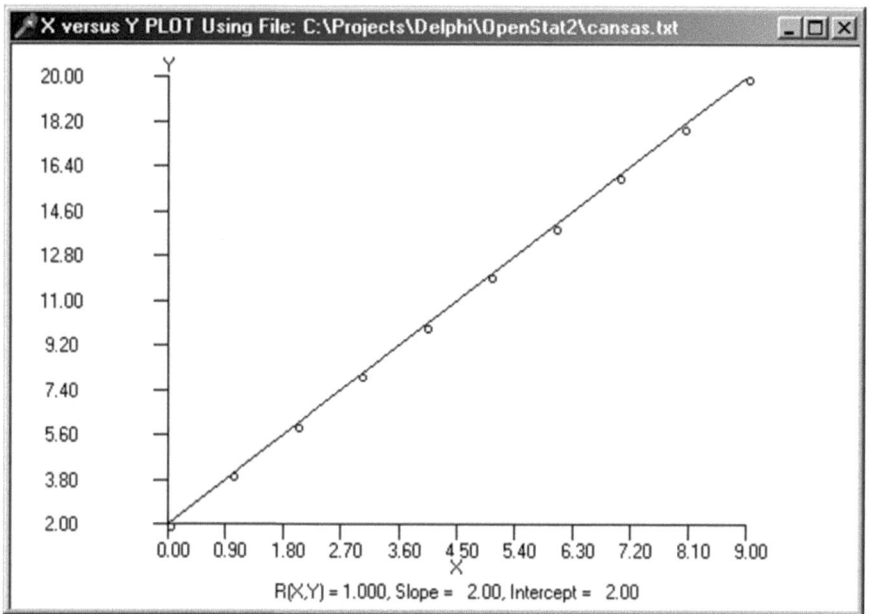

Fig. 3.7 Plot for a correlation of 1.0

and

$$z_y = (Y_i - \bar{Y})/\sigma_y \tag{3.14}$$

then we may write for our prediction of the corresponding zy scores

$$z_y' = bz_x + 0 \tag{3.15}$$

since the intercept is zero for z scores.

The Least-Squares criterion implies that the squared difference between each predicted score and actual observed score Y is a minimum. That is

$$\sum_{i=1}^{N}(z_y - z_y')^2 = \text{Minimum} \tag{3.16}$$

where z_y' is the predicted z_y score for an individual.

The problem is to obtain values of b such that the above statement is true. If we substitute bz_x for each z_y' in the above equation and expand, we get

$$\text{Min} = \sum_{i=1}^{N}[z_y - bz_x]^2 \tag{3.17}$$

$$= \sum_{i=1}^{N} \left(z_y{}^2 + b^2 z_x{}^2 - 2b z_y z_x\right)$$

$$= \sum_{i=1}^{N} z_y{}^2 + b2 \sum_{i=1}^{N} z_x{}^2 - 2b \sum_{i=1}^{N} z_y z_x \qquad (3.18)$$

In the mathematics called Calculus, it is learned that the first derivative of a function is either a minimum or a maximum. By taking the partial derivative of the above function Min (we will call it M) with respect to b, we get an equation which can be solved for b. This equation is set equal to zero and solved for b. The derivative of M with respect to be is:

$$\frac{\delta M}{\delta b} = 2b \, \Sigma \, z_x{}^2 - 2\Sigma \, z_y z_x \qquad (3.19)$$

Setting the derivative to zero and solving for b gives

$$0 = b \, \Sigma \, z_x{}^2 - \Sigma \, z_y z_x \qquad (3.20)$$

$$\text{or } b = \Sigma \, z_y z_x / \Sigma \, z_x{}^2 \qquad (3.21)$$

Since the sum of squared z scores is equal to N (if we use the biased standard deviation), we see that

$$b = \Sigma \, z_x z_y / N.$$

The product–moment correlation was earlier defined to be the average of z score products. Therefore, the slope of a regression line in z score form is simply

$$b = r_{xy}$$

The prediction equation is therefore

$$z_y{}' = r_{xy} z_x \qquad (3.21)$$

To determine the values of B and C in the equation for raw scores, simply substitute the definition of z scores in the above equation, that is

$$\frac{(Y' - \bar{Y})}{s_y} = r_{xy} \frac{(X - \bar{X})}{s_x} \qquad (3.22)$$

$$\text{or } (Y' - \bar{Y}) = r_{xy} \frac{s_y}{s_x} (X - \bar{X}) \qquad (3.23)$$

The Variance of Predicted Scores

$$\text{or} \quad Y' = r_{xy}\frac{s_y}{s_x}X - r_{xy}\frac{s_y}{s_x}\bar{X} + \bar{Y} \tag{3.24}$$

Letting $B = r_{xy}(s_y/s_x)$, the last equation may be written

$$Y' = BX - (B\bar{X} - \bar{Y}) \tag{3.25}$$

To express the equation is the typical "straight line" equation, let

$$C = \bar{Y} - B\bar{X} \tag{3.26}$$

so that

$$Y' = BX + C \tag{3.27}$$

To summarize, the least-squares criterion is met when the predicted scores for z_y or Y are obtained from

$$z_y' = r\, z_x \tag{3.28}$$

or

$$Y' = BX + C \text{ where } B = r_{xy}(s_y/s_x) \text{ and}$$

$$C = \bar{Y} - B\bar{X}$$

The Variance of Predicted Scores

We can develop an expression for the variance of predicted scores z_y' or Y'. Using the definition of variance, we have

$$s^2Y' = \frac{(Y' - \bar{Y})^2}{N} \tag{3.29}$$

By substituting the definition of Y', that is, $BX + C$, in the above equation, we could show that the variance of predicted scores is

$$s^2Y' = r_{xy}^2 s_y^2 \tag{3.30}$$

That is, the variance of the predicted scores is the square of the product–moment correlation between X and Y times the variance of the Y scores. It is also useful to rewrite the above equation as

$$r_{xy}^2 = s^2 Y' / s_y^2. \tag{3.31}$$

The square of the correlation is that proportion of total score variance that is predicted by X !

The Variance of Errors of Prediction

Just as we developed an expression for the variance of predicted scores above, we can also develop an expression for the variance of errors of prediction, that is, the variance of $e_i = (Y_i - Y_i')$ for each score.

Again using the definition of variance we can write

$$s^2 Y.X = \frac{\Sigma (Y_i - Y_i')^2}{N} \tag{3.32}$$

This formula is biased due to estimating both the mean of X as well as the mean of Y in the population. For that reason the unbiased estimate is

$$s^2 Y.X = \frac{\Sigma e_i^2}{N - 2} \tag{3.33}$$

The square root of this variance is called the **standard error of estimate**. When we can assume the errors of prediction are normally distributed, it allows us to estimate a confidence interval for a given predicted score.

Rather than having to compute an error for each individual, the above formula may be translated into a more convenient computational form:

$$s^2 Y.X = s_y^2 (1 - r_{xy}^2) \frac{N - 1}{N - 2} \tag{3.34}$$

As an example in using the standard error of estimate, assume we have obtained a correlation of 0.8 between scores of X and Y for 40 subjects. If the variance of the Y scores is 100, then the variance of estimate is

$$s^2 Y.X = 100 \ (1.0 - 0.64) \ (19 \ / \ 18)$$
$$= 38$$

and

$$S_{Y.X} = \sqrt{38} = 6.1644$$

Using plus or minus one under the normal distribution, we can state that a predicted Y score would be expected to be in the interval (Y' ± 6.2) approximately 68% of the time.

Testing Hypotheses Concerning the Pearson Product–Moment Correlation

Hypotheses About Correlations in One Population

The product–moment correlation is an index of the linear relationship between two variables that varies between -1.0 and $+1.0$ with a value of 0.0 indicating no relationship. When obtaining pairs of X and Y scores on a sample of subjects drawn from a population, one can hypothesize that the correlation in the population does not differ from zero (0), i.e. Ho: $\mu = 0$. The test statistic is:

$$t = \frac{r - \gamma}{S_r} \text{ with } n - 2 \text{ degrees of freedom, and} \quad (3.35)$$

$$S_r = \sqrt{\frac{1 - r^2}{n - 2}} \quad (3.36)$$

As an example, assume a sample correlation $r = 0.3$ is obtained from a random selection of 38 subjects from a population of subjects. To test the hypothesis that the population correlation does not differ significantly from zero in either direction, we would obtain

$$S_r = \sqrt{\frac{1 - .09}{38 - 2}} = 0.158989866 \quad (3.37)$$

and

$$t = r / S_r = .3 / 0.158989866 = 1.886912706 \quad (3.38)$$

With $n - 2 = 36$ degrees of freedom, the t value obtained would be considered significant at the 0.05 level for a one-tailed test ($r > 0$), hence we would fail to retain the null hypothesis (reject).

Test That the Correlation Equals a Specific Value

The sampling distribution of the product–moment correlation is approximately normal or t distributed when sampled from a population in which the true correlation is zero. Occasionally, however, one wishes to test the hypothesis that the population correlation does not differ from some specified value ρ not equal to zero. The distribution of sample correlations from a population in which the correlation differs from zero is skewed, with the degree of skewness increasing as the population correlation differs from zero. It is possible to transform the correlations to a statistic which has a sampling distribution that is approximately normal in shape. The transformation, credited to Fisher, is:

$$z_r = 0.5 \log_e \left(\frac{1+r}{1-r} \right) \qquad (3.39)$$

This statistic has a standard error of:

$$S_r = \sigma[1/(n-3)] \qquad (3.40)$$

Using the above, a *t*-test for the hypothesis Ho:ρ = a can be obtained as

$$z = \frac{z_r - z_\rho}{S_r} \qquad (3.41)$$

For example, assume we have obtained a sample correlation of r = 0.6 on 50 subjects and we wish to test the hypothesis that the population correlation does not differ from 0.5 in the positive direction. We would first transform both the sample and population correlations to the Fisher's z score and obtain:

$$z_r = .5\log_e[(1+.6)/(1-.6)] = 0.6931472 \qquad (3.42)$$

and

$$z_\rho = .5\log_e[(1+.5)/(1-.5)] = 0.5493061 \qquad (3.43)$$

Next, we obtain the standard error as

$$S_{z_r} = \sigma[1/(n-3)] = \sigma[1/(50-3)] = 0.145865 \qquad (3.44)$$

Our test statistic is then

$$z = \frac{z_r - z_\rho}{S_{z_r}} = \frac{0.143841}{0.145865} = 0.986 \qquad (3.45)$$

Test That the Correlation Equals a Specific Value

Approximately .16 of the area of the normal curve lies beyond a z of .986. We would retain our null hypothesis if our decision rule was for a probability of 0.05 or less in order to reject.

As for all of the sample statistics discussed so far, a confidence interval may be constructed. In the case of the Fisher's z transformation of the correlation, we first construct our interval using the z-transformed scores and then obtain the anti-log to express the interval in terms of product–moment correlations. For example, the 90% Confidence Interval for the above data is obtained as:

$$\begin{aligned} CI_{90} &= z_r \pm 1.645(S_{z_r}) \\ &= .693 \pm 1.645(.146) = .693 \pm 0.24 \\ &= (.453, .933) \end{aligned} \qquad (3.46)$$

and transforming the z_r intervals to r intervals gives

$$CI_{90} = (0.424, 0.732) \qquad (3.47)$$

We converted the z_r values back to correlations using

$$r = \frac{e^{2z_r} - 1}{e^{2z_r} + 1} \qquad (3.48)$$

Notice that the sample value of 0.6 is "captured" in the 90% Confidence Interval, thus verifying our one-tailed 0.05 test.

OpenStat contains a procedure for completing a z test for data like that presented above.

Under the Statistics menu, move your mouse down to the Comparisons sub-menu, and then to the option entitled "One Sample Tests". When the form below displays, click on the Correlation button and enter the sample value .5, the population value .6, and the sample size 50. Change the confidence level to 90.0% (Fig. 3.8).

Shown below is the z-test for the above data:
ANALYSIS OF A SAMPLE CORRELATION

```
Sample Correlation = 0.600
Population Correlation = 0.500
Sample Size = 50
z Transform of sample correlation = 0.693
z Transform of population correlation = 0.549
Standard error of transform = 0.146
z test statistic = 0.986 with probability 0.838
z value required for rejection = 1.645
Confidence Interval for sample correlation = (0.425, 0.732)
```

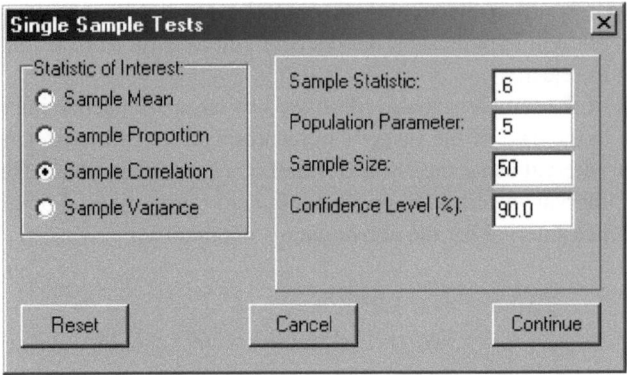

Fig. 3.8 Single sample tests dialog form

Testing Equality of Correlations in Two Populations

When two populations have been sampled, a correlation between X and Y scores of each sample are often obtained. We may test the hypothesis that the product–moment correlation in the two populations are equal. If we assume the samples are independent, our test statistic may be obtained as

$$z = \frac{(z_{r_1} - z_{r_2}) - (z_{\gamma_1} - z_{\gamma_2})}{S_{(z_{r_1} - z_{r_2})}} \tag{3.49}$$

where

$$S_{(z_{r_1} - z_{r_2})} = \sqrt{\frac{1}{n_1 - 3} + \frac{1}{n_2 - 3}} \tag{3.50}$$

As an example, assume we have collected ACT Composite scores (a college aptitude test) and College Freshman Grade Point Average (GPA) scores for both men and women at a state university. We might hypothesize that in the population of men and women at this university, there is no difference between the correlation of GPA and ACT. Now pretend that a sample of 30 men yielded a correlation of .5 and that a sample of 40 women yielded a correlation of .6. Our test would yield:

$$z_r = 0.5493061 \text{ for the men,}$$

$$z_r = 0.6931472 \text{ for the women, and}$$

$$S_{(z_{r_1}-z_{r_2})} = \sqrt{\frac{1}{27}+\frac{1}{37}} = 0.253108798 \qquad (3.51)$$

and the test value of

$$z = (0.5493061 - 0.6931472)/0.253108798$$
$$= -0.568$$

which would not be significant.

The above test reflects the use of Fisher's log transformation of a correlation coefficient to an approximate z score. The correlations in each sample are converted to z's and a test of the difference between the z scores is performed. In this case, the difference obtained had a relatively large chance of occurrence when the null hypothesis is true (0.285) and the 95% confidence limit brackets the sample difference of 0.253. The Fisher z transformation of a correlation coefficient is

$$z_r = \frac{1}{2}\log_e\left(\frac{1+r}{1-r}\right) \qquad (3.52)$$

The test statistic for the difference between the two correlations is:

$$z_r = \frac{(z_{r_1}-z_{r_2})-(z_{\rho_1}-z_{\rho_2})}{\sigma_{(z_{r_1}-z_{r_2})}} \qquad (3.53)$$

where the denominator is the standard error of difference between two independent transformed correlations:

$$\sigma_{(z_{r_1}-z_{r_2})} = \sqrt{\left(\frac{1}{n_1-3}\right)\left(\frac{1}{n_2-3}\right)} \qquad (3.54)$$

The confidence interval is constructed for the difference between the obtained z scores and the interval limits are then translated back to correlations. The confidence limit for the z scores is obtained as:

$$CI_\% = (z_{r_1}-z_{r_2})+/-z_\%\sigma_{(z_{r_1}-z_{r_2})} \qquad (3.55)$$

We can then translate the obtained upper and lower z values using:

$$r = \frac{e^{2z_r}-1}{e^{2z_r}+1} \qquad (3.56)$$

For the test that two dependent correlations do not differ from zero we use the following *t*-test:

$$t = \frac{(r_{xy} - r_{xz})\sqrt{(n-3)(1+r_{yz})}}{\sqrt{2\left(1 - r_{xy}^2 - r_{xz}^2 - r_{yz}^2 + 2r_{xy}r_{xz}r_{yz}\right)}} \qquad (3.57)$$

We would therefore conclude that, in the populations sampled, there is not a significant difference between the correlations for men and women. Using OpenStat to accomplish the above calculations is rather easy. Under the Statistics menu move to the Comparisons sub-menu and further in that menu to the Two-Sample Tests sub-sub-menu. Click on the Independent Correlations option. Shown below are the results for the above data:

```
COMPARISON OF TWO CORRELATIONS

Correlation one = 0.500
Sample size one = 30
Correlation two = 0.600
Sample size two = 40
Difference between correlations = -0.100
Confidence level selected = 95.0
z for Correlation One = 0.549
z for Correlation Two = 0.693
z difference = -0.144
Standard error of difference = 0.253
z test statistic = -0.568
Probability > |z| = 0.715
z Required for significance = 1.960
Note: above is a two-tailed test.
Confidence Limits = (-0.565, 0.338)
```

Differences Between Correlations in Dependent Samples

Assume that three variables are available for a population of subjects. For example, you may have ACT scores, Freshman GPA (FGPA) scores and High School GPA (HSGPA) scores. It may be of interest to know whether the correlation of ACT scores with High School GPA is equal to the correlation of ACT scores with the Freshman GPA obtained in College. Since the correlations would be obtained across the same subjects, we have dependency between the correlations. In other words, to test the hypothesis that the two correlations r_{xy} and r_{xz} are equal, we must take into consideration the correlation r_{yz}. A t-test with degrees of freedom equal to N-3 may be obtained to test the hypothesis that $\mu_{xy} = \mu_{xz}$ in the population. Our t-test is constructed as

$$t = \frac{r_{x,y} - r_{x,z}}{\sqrt{\frac{2\left(1 - r_{x,y}^2 - r_{x,z}^2 - r_{y,z}^2 + 2r_{x,y}r_{x,z}r_{y,z}\right)}{(N-3)(1+r_{y,z})}}} \qquad (3.58)$$

Assume we have drawn a sample of 50 college freshman and observed:

$r_{xy} = .4$ for the correlation of ACT and FGPA, and
$r_{xz} = .6$ for the correlation of ACT and HSGPA, and
$r_{yz} = .7$ for the correlation of FGPA and HSGPA.

Then for the hypothesis that $\mu_{xy} = \mu_{xz}$ in the population of students sampled, we have

$$t = \frac{.4 - .6}{\sqrt{\frac{2[1 - .4^2 - .6^2 - .7^2 + 2(.4)(.6)(.7)]}{(50 - 3)(1 + .7)}}} = \frac{-.2}{\sqrt{\frac{.652}{79.9}}} = \frac{-.2}{0.0903338}$$

$$= 2.214 \qquad (3.59)$$

This sample t value has a two-tailed probability of less than 0.05. If the 0.05 level were used for our decision process, we would reject the hypothesis of equal correlations of ACT with the high school GPA and the freshman college GPA. It would appear that the correlation of the ACT with high school GPA is greater than with College GPA in the population studied.

Again, OpenStat provides the computations for the difference between dependent correlations as shown in the figure below (Fig. 3.9):

```
COMPARISON OF TWO CORRELATIONS
Correlation x with y = 0.400
Correlation x with z = 0.600
Correlation y with z = 0.700
Sample size = 50
Confidence Level Selected = 95.0
Difference r(x,y) - r(x,z) = -0.200
t test statistic = -2.214
Probability > |t| = 0.032
t value for significance = 2.012
```

Partial and Semi-Partial Correlations

What did one regression coefficient say to the other regression coefficient? I'm partial to you!

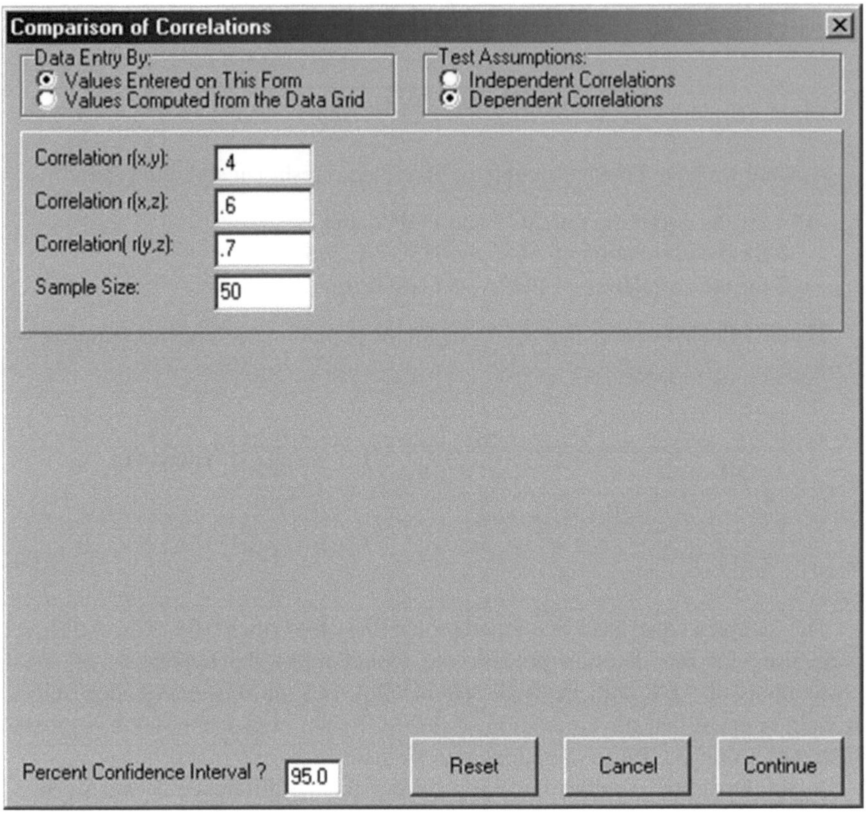

Fig. 3.9 Form for comparison of correlations

Partial Correlation

One is often interested in knowing what the product–moment correlation would be between two variables if one or more related variables could be held constant. For example, in one of our previous examples, we may be curious to know what the correlation between achievements in learning French is with past achievement in learning English with intelligence held constant. In other words, if that proportion of variance shared by both French and English learning with IQ is removed, what is the remaining variance shared by English and French?

When one subtracts the contribution of a variable, say, X_3, from both variables of a correlation say, X_1 and X_2, we call the result the partial correlation of X_1 with X_2 partialling out X_3. Symbolically this is written as $r_{12.3}$ and may be computed by

Partial and Semi-Partial Correlations

$$r_{12.3} = \frac{r_{12} - r_{13}r_{23}}{\sqrt{(1-r_{13}^2)(1-r_{23}^2)}} \tag{3.60}$$

More than one variable may be partialled from two variables. For example, we may wish to know the correlation between English and French achievement partialling both IQ and previous Grade Point Average. A general formula for multiple partial correlation is given by

$$r_{12.34..k} = \frac{(1.0 - R^2_{y.34..k}) - (1.0 - R^2_{y.12..k})}{1.0 - R^2_{y.34..k}} \tag{3.61}$$

Semi-Partial Correlation

It is not necessary to partial out the variance of a third variable from both variables of a correlation. It may be the interest of the researcher to partial a third variable from only one of the other variables. For example, the researcher in our previous example may feel that intelligence should be left in the variance of the past English achievement which has occurred over a period of years but should be removed from the French achievement which is a much shorter learning experience. When the variance of a third variable is partialled from only one of the variables in a correlation, we call the result a semi-partial or part correlation. The symbol and calculation of the part correlation is

$$r_{1(2.3)} = \frac{r_{1,2} - r_{1,3}r_{2,3}}{\sqrt{(1.0 - r^2_{23})}} \tag{3.62}$$

where X_3 is partialled only from X_2.

The squared multiple correlation coefficient R^2 may also be expressed in terms of semi_partial correlations. For example, we may write the equation

$$R^2_{y.1\,2\cdot\cdot k} = r^2_{y.1} + r^2_{y(2.1)} + r^2_{y(3.12)} + .. + r^2_{y(k.12..k-1)} \tag{3.63}$$

In this formula, each semi-partial correlation reflects the proportion of variance contributed by a variable independent of previous variables already entered in the equation. However, the order of entry is important. Any given variable may explain a different proportion of variance of the independent variable when entered first, say, rather than last!

The semi-partial correlation of two variables in which the effects of K-1 other variables have been partialed from the second variable may be obtained by multiple regression. That is

$$r^2_{y(1.23\ ..\ k)} = R^2_{2y.1\ 2\ ..\ k} - R^2_{y.23..k} \tag{3.64}$$

Autocorrelation

A large number of measurements are collected over a period of time. Stock prices, quantities sold, student enrollments, grade point averages, etc. may vary systematically across time periods. Variations may reflect trends which repeat by week, month or year. For example, a grocery item may sell at a fairly steady rate on Tuesday through Thursday but increase or decrease on Friday, Saturday, Sunday and Monday. If we were examining product sales variations for a product across the days of a year, we might calculate the correlation between units sold over consecutive days. The data might be recorded simply as a series such as "units sold" each day. The observations can be recorded across the columns of a grid or as a column of data in a grid. As an example, the grid might contain:

CASE/VAR	Day	Sold
Case 1	1	34
Case 2	2	26
Case 3	3	32
Case 4	4	39
Case 5	5	29
Case 6	6	14
...		
Case 216	6	15
Case 217	7	12

If we were to copy the data in the above "Sold" column into an adjacent column but starting with the Case 2 data, we would end up with:

CASE/VAR	Day	Sold	Sold2
Case 1	1	34	26
Case 2	2	26	32
Case 3	3	32	39
Case 4	4	39	29
Case 5	5	29	14
Case 6	6	14	11
...			
Case 216	6	15	12
Case 217	7	12	—

In other words, we repeat our original scores from Case 2 through case 217 in the second column but moved up one row. Of course, we now have one fewer case with complete data in the second column. We say that the second column of data "lags" the first column by 1. In a similar fashion we might create a third, fourth, fifth, etc. column representing lags of 2, 3, 4, 5, etc.. Creating lag variables 1 through 6 would result in variables starting with sales on days 1 through 7, that is, a week of sale data. If we obtain the product–moment correlations for these seven variables, we would have the correlations among Monday sales, Tuesday Sales, Wednesday Sales, etc. We note that the mean and variance are best estimated by the lag 0 (first column) data since it contains all of the observations (each lag loses one additional observation.) If the sales from day to day represent "noise" or simply random variations then we would expect the correlations to be close to zero. If, on the other hand, we see an systematic increase or decrease in sales between say, Monday and Tuesday, then we would observe a positive or negative correlation.

In addition to the inter-correlations among the lagged variables, we would likely want to plot the average sales for each. Of course, these averages may reflect simply random variation from day to day. We may want to "smooth" these averages to enhance our ability to discern possible trends. For example, we might want the average of day 3 to be a weighted average of that day plus the previous 2 day sales. This "moving average" would tend to smooth random peaks and valleys that occur from day to day.

It is also the case that an investigator may want to predict the sales for a particular day based on the previous sales history. For example, we may want to predict day 8 sales given the history of previous 7 day sales.

Now let us look at an example of auto-correlation. We will use a file named strikes.tab. The file contains a column of values representing the number of strikes which occurred each month over a 30 month period. Select the auto-correlation procedure from the Correlations sub-menu of the Statistics main menu. Below is a representation of the form as completed to obtain auto-correlations, partial auto-correlations, and data smoothing using both moving average smoothing and polynomial regression smoothing (Fig. 3.10):

When we click the Compute button, we first obtain a dialog form for setting the parameters of our moving average.

In that form we first enter the number of values to include in the average from both sides of the current average value. We selected two. Be sure and press the Enter key after entering the order value. When you do, two theta values will appear in a list box. When you click on each of those thetas, you will see a default value appear in a text box. This is the weight to assign the leading and trailing averages (first or second in our example.) In our example we have accepted the default value for both thetas (simply press the Return key to accept the default or enter a value and press the Return key.) Now press the Apply button. When you do this, the weights for all of the values (the current mean and the 1, 2, ... order means) are recalculated. You can then press the OK button to proceed with the process (Fig. 3.11).

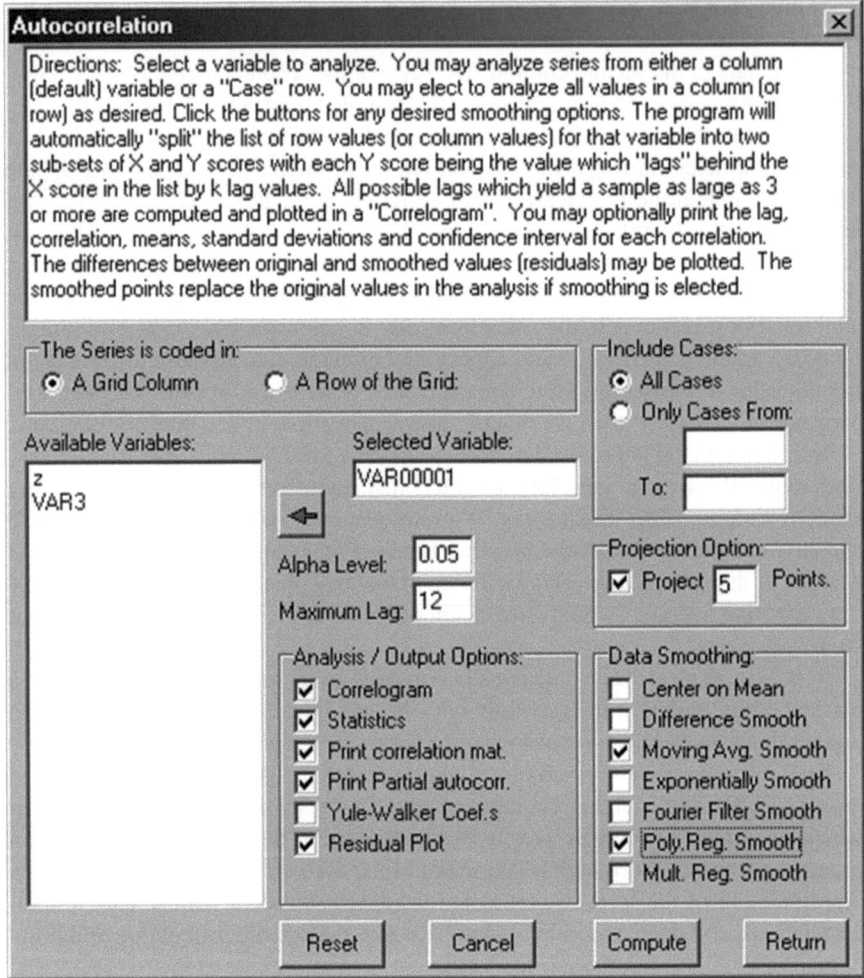

Fig. 3.10 The autocorrelation dialog

The procedure then plots the original (30) data points and their moving average smoothed values. Since we also asked for a projection of five points, they too are plotted. The plot should look like that shown below (Fig. 3.12):

We notice that there seems to be a "wave" type of trend with a half-cycle of about 15 months. When we press the Return button on the plot of points we next get the following (Fig. 3.13):

This plot shows the original points and the difference (residual) of the smoothed values from the original. At this point, the procedure replaces the original points

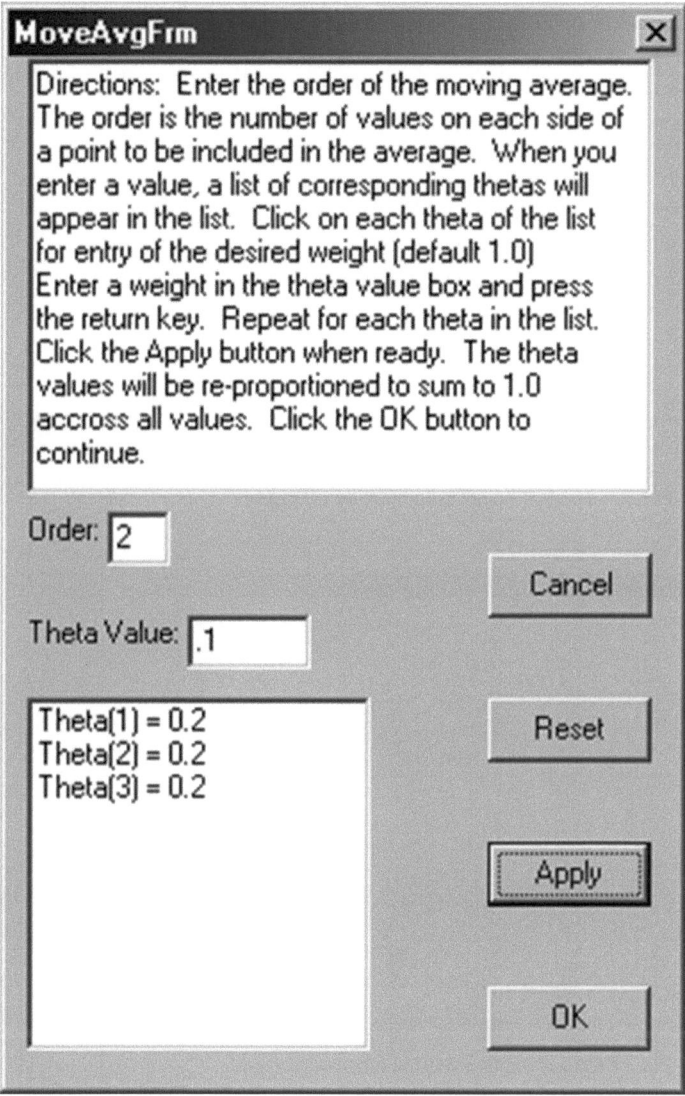

Fig. 3.11 The moving average dialog

with the smoothed values. Press the Return button and you next obtain the following (Fig. 3.14):

This is the form for specifying our next smoothing choice, the polynomial regression smoothing. We have elected to use a polynomial value of 2 which will result in a model for a data point $Y_{t-1} = B * t^2 + C$ for each data point. Click the OK button to proceed. You then obtain the following result (Fig. 3.15):

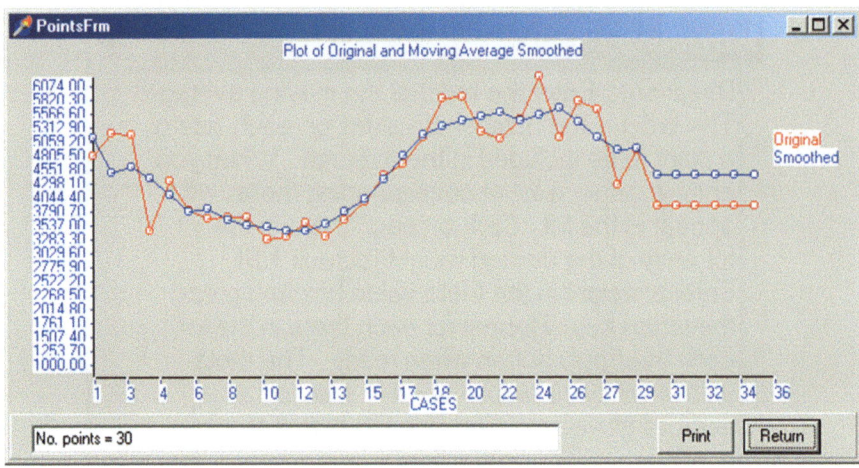

Fig. 3.12 Plot of smoothed points using moving averages

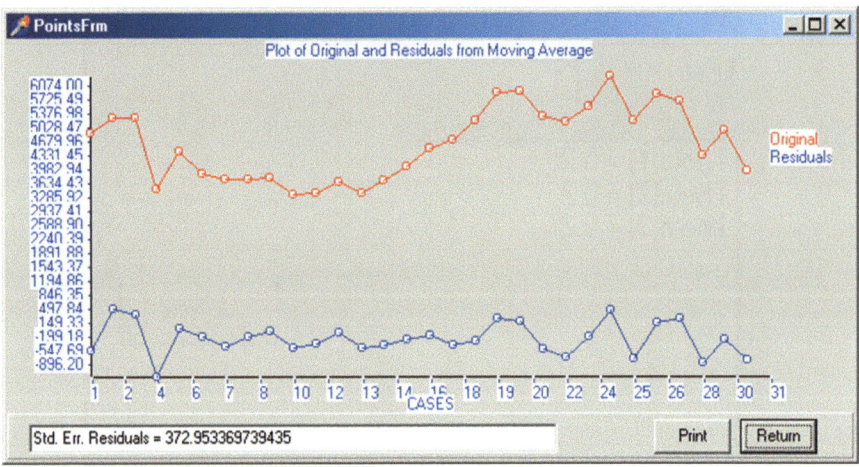

Fig. 3.13 Plot of residuals obtained using moving averages

It appears that the use of the second order polynomial has "removed" the cyclic trend we saw in the previously smoothed data points. Click the return key to obtain the next output as shown below (Fig. 3.16):

This result shows the previously smoothed data points and the residuals obtained by subtracting the polynomial smoothed points from those previous points. Click the Return key again to see the next output shown below:

Autocorrelation

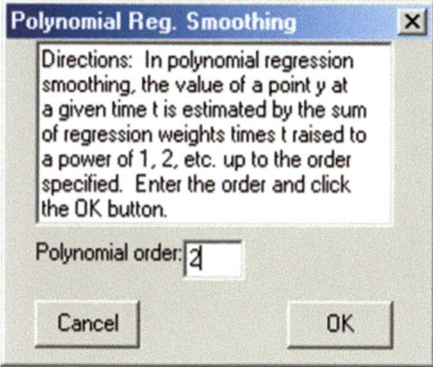

Fig. 3.14 Polynomial regression smoothing form

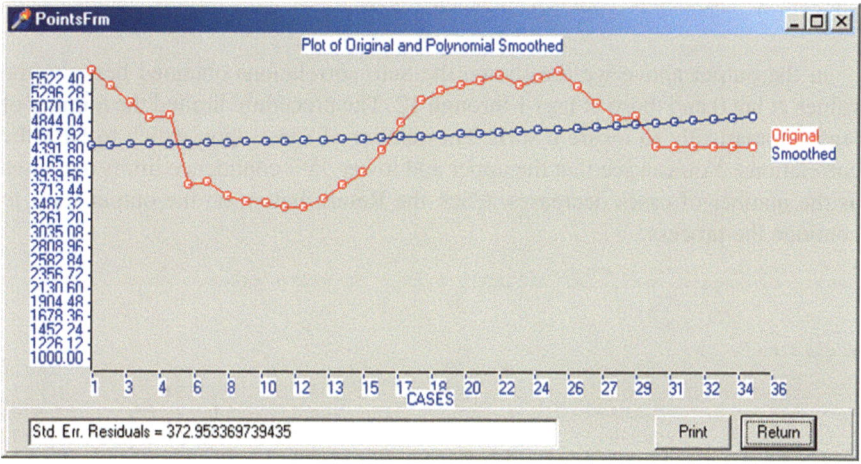

Fig. 3.15 Plot of polynomial smoothed points

```
Overall mean = 4532.604, variance = 11487.241
Lag      Rxy      MeanX      MeanY      Std.Dev.X  Std.Dev.Y   Cases      LCL        UCL

  0    1.0000   4532.6037  4532.6037   109.0108   109.0108      30      1.0000     1.0000
  1    0.8979   4525.1922  4537.3814   102.9611   107.6964      29      0.7948     0.9507
  2    0.7964   4517.9688  4542.3472    97.0795   106.2379      28      0.6116     0.8988
  3    0.6958   4510.9335  4547.5011    91.3660   104.6337      27      0.4478     0.8444
  4    0.5967   4504.0864  4552.8432    85.8206   102.8825      26      0.3012     0.7877
  5    0.4996   4497.4274  4558.3734    80.4432   100.9829      25      0.1700     0.7287
  6    0.4050   4490.9565  4564.0917    75.2340    98.9337      24      0.0524     0.6679
  7    0.3134   4484.6738  4569.9982    70.1928    96.7340      23     -0.0528     0.6053
  8    0.2252   4478.5792  4576.0928    65.3196    94.3825      22     -0.1470     0.5416
  9    0.1410   4472.6727  4582.3755    60.6144    91.8784      21     -0.2310     0.4770
 10    0.0611   4466.9544  4588.8464    56.0772    89.2207      20     -0.3059     0.4123
 11   -0.0139   4461.4242  4595.5054    51.7079    86.4087      19     -0.3723     0.3481
 12   -0.0836   4456.0821  4602.3525    47.5065    83.4415      18     -0.4309     0.2852
```

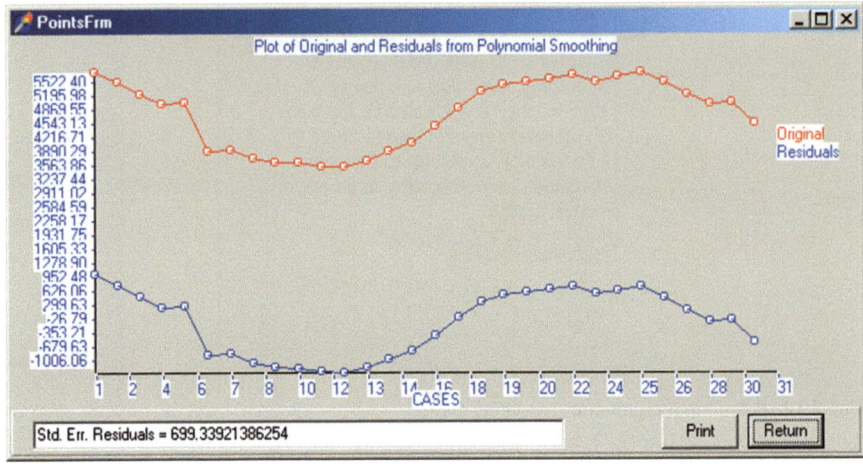

Fig. 3.16 Plot of residuals from polynomial smoothing

In the output above we are shown the auto-correlations obtained between the values at lag 0 and those at lags 1 through 12. The procedure limited the number of lags automatically to insure a sufficient number of cases upon which to base the correlations. You can see that the upper and lower 95% confidence limits increases as the number of cases decreases. Click the Return button on the output form to continue the process.

```
Matrix of Lagged Variable: VAR00001 with   30 valid cases.


Variables
                Lag 0       Lag 1       Lag 2       Lag 3       Lag 4
     Lag  0     1.000       0.898       0.796       0.696       0.597
     Lag  1     0.898       1.000       0.898       0.796       0.696
     Lag  2     0.796       0.898       1.000       0.898       0.796
     Lag  3     0.696       0.796       0.898       1.000       0.898
     Lag  4     0.597       0.696       0.796       0.898       1.000
     Lag  5     0.500       0.597       0.696       0.796       0.898
     Lag  6     0.405       0.500       0.597       0.696       0.796
     Lag  7     0.313       0.405       0.500       0.597       0.696
     Lag  8     0.225       0.313       0.405       0.500       0.597
     Lag  9     0.141       0.225       0.313       0.405       0.500
     Lag 10     0.061       0.141       0.225       0.313       0.405
     Lag 11    -0.014       0.061       0.141       0.225       0.313
     Lag 12    -0.084      -0.014       0.061       0.141       0.225
```

Autocorrelation

```
Partial Correlation Coefficients with    30 valid cases.

Variables        Lag 0         Lag 1         Lag 2         Lag 3
Lag 4
                 1.000         0.898         -0.051        -0.051        -
0.052

Variables        Lag 5         Lag 6         Lag 7         Lag 8
Lag 9
                 -0.052        -0.052        -0.052        -0.052        -
0.051

Variables        Lag 10        Lag 11
                 -0.051        -0.051
```

The above data presents the inter-correlations among the 12 lag variables. Click the output form's Return button to obtain the next output:

```
Variables
                Lag 5         Lag 6         Lag 7         Lag 8         Lag 9
     Lag 0      0.500         0.405         0.313         0.225         0.141
     Lag 1      0.597         0.500         0.405         0.313         0.225
     Lag 2      0.696         0.597         0.500         0.405         0.313
     Lag 3      0.796         0.696         0.597         0.500         0.405
     Lag 4      0.898         0.796         0.696         0.597         0.500
     Lag 5      1.000         0.898         0.796         0.696         0.597
     Lag 6      0.898         1.000         0.898         0.796         0.696
     Lag 7      0.796         0.898         1.000         0.898         0.796
     Lag 8      0.696         0.796         0.898         1.000         0.898
     Lag 9      0.597         0.696         0.796         0.898         1.000
     Lag 10     0.500         0.597         0.696         0.796         0.898
     Lag 11     0.405         0.500         0.597         0.696         0.796
     Lag 12     0.313         0.405         0.500         0.597         0.696

Variables
                Lag 10        Lag 11        Lag 12
     Lag 0      0.061         -0.014        -0.084
     Lag 1      0.141         0.061         -0.014
     Lag 2      0.225         0.141         0.061
     Lag 3      0.313         0.225         0.141
     Lag 4      0.405         0.313         0.225
     Lag 5      0.500         0.405         0.313
     Lag 6      0.597         0.500         0.405
     Lag 7      0.696         0.597         0.500
     Lag 8      0.796         0.696         0.597
     Lag 9      0.898         0.796         0.696
     Lag 10     1.000         0.898         0.796
     Lag 11     0.898         1.000         0.898
     Lag 12     0.796         0.898         1.000
```

Variables	Lag 0	Lag 1	Lag 2	Lag 3	Lag 4
					0.500
Lag 10	0.061	0.141	0.225	0.313	0.405
Lag 11	-0.014	0.061	0.141	0.225	0.313
Lag 12	-0.084	-0.014	0.061	0.141	0.225

Variables	Lag 5	Lag 6	Lag 7	Lag 8	Lag 9
Lag 0	0.500	0.405	0.313	0.225	0.141
Lag 1	0.597	0.500	0.405	0.313	0.225
Lag 2	0.696	0.597	0.500	0.405	0.313
Lag 3	0.796	0.696	0.597	0.500	0.405
Lag 4	0.898	0.796	0.696	0.597	0.500
Lag 5	1.000	0.898	0.796	0.696	0.597
Lag 6	0.898	1.000	0.898	0.796	0.696
Lag 7	0.796	0.898	1.000	0.898	0.796
Lag 8	0.696	0.796	0.898	1.000	0.898
Lag 9	0.597	0.696	0.796	0.898	1.000

The partial auto-correlation coefficients represent the correlation between lag 0 and each remaining lag with previous lag values partialled out. For example, for lag 2 the correlation of −0.051 represents the correlation between lag 0 and lag 2 with lag 1 effects removed. Since the original correlation was 0.796, removing the effect of lag 1 made a considerable impact. Again click the Return button on the output form. Next you should see the following results (Fig. 3.17):

This plot or "correlogram" shows the auto-correlations and partial auto-correlations obtained in the analysis. If only "noise" were present, the correlations would vary around zero. The presence of large values is indicative of trends in the data.

Series

Introduction

In many areas of research observations are taken periodically of the same object. For example, a medical researcher may take hourly blood pressure readings of a patient. An economist may record the price of a given stock each day for a long

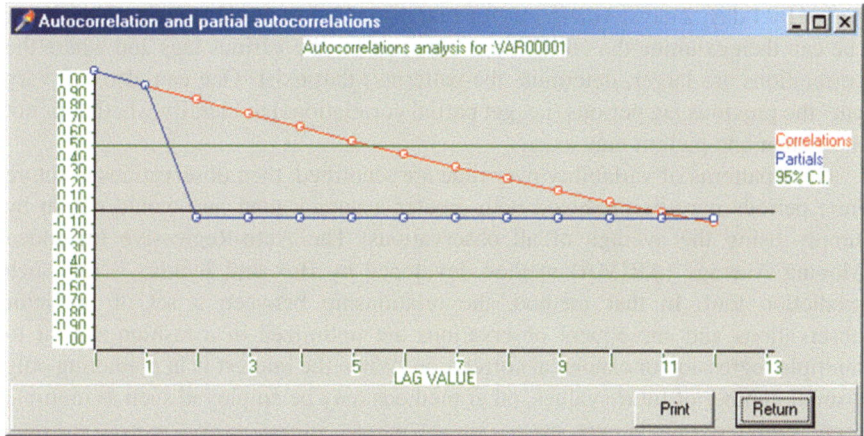

Fig. 3.17 Auto and partial autocorrelation plot

period. A retailer may record the number of units sold of a particular item on a daily basis. An industrialist may record the number of parts rejected each day over a period of time. In each of these cases, the researcher may be interested in identifying patterns in the fluctuation of the observations. For example, does a patient's systolic blood pressure systematically increase or decrease during visits by relatives? Do stock prices tend to vary systematically from month to month? Does the number of cans of tomato soup sold vary systematically across the days of the week or the months? Does the number of parts rejected in the assembly line vary systematically with the time of day or day of the week?

One approach often taken to discern patterns in repeated measurements is to simply plot the observed values across the time intervals on which the recording took place. This may work well to identify major patterns in the data. Sometimes however, factors which contribute to large systematic variations may "hide" other patterns that exist. A variety of methods have been developed to identify such patterns. For example, if the patterns are thought to potentially follow a sin wave pattern across time, a Fourier analysis may be used. This method takes a "signal" such as an electrical signal or a series of observations such as units sold each day and attempts to decompose the signal into fundamental frequencies. Knowing the frequencies allows the researcher to identify the "period" of the waves. Another method often employed involves examining the product–moment correlation between observations beginning at a specific "lag" period from each other. For example, the retailer may create an "X" variable beginning on a Monday and and "Y" variable beginning on the Monday 4 weeks later. The number of units sold are then recorded for each of these Mondays, Tuesdays, etc. If there is a systematic variation in the number of units sold over the weeks of this lag, the correlation will tend to be different from zero. If, on the other hand, there is only random variation, the correlation would be expected to be zero. In fact, the retailer may vary the lag period by 1 day, 2 days, 3 days, etc. for a large number of possible lag periods. He or she can then examine the correlations obtained for the various lags and where the correlations are larger, determine the pattern(s) that exist. One can also "co-vary out" the previous lag periods (i.e. get partial correlations) to identify whether or not more than one pattern may exist.

Once patterns of variability over time are identified, then observations at future time periods may be predicted with greater accuracy than one would obtain by simply using the average of all observations. The Auto-Regressive Imbedded Moving Average (ARIMA) method developed by Box and Jenkins is one such prediction tool. In that method, the relationship between a set of predictor observations and subsequent observations are optimized in a fashion similar to multiple regression or canonical correlation. When the interest is in predicting only a small number of future values, other methods may be employed such as multiple regression, moving average, etc.

The OpenStat program provides the means for obtaining auto-correlations, partial auto-correlations, Fourier analysis, moving average analysis and other tools useful for time series analyses.

Chapter 4
Multiple Regression

> *This chapter develops the theory and applications of Multiple Linear Regression Analysis. The multiple regression methods are frequently used (and misused.) It also forms the heart of several other analytic methods including Path Analysis, Structural Equation Modeling and Factor Analysis.*

The Linear Regression Equation

One of the major applications in statistics is the prediction of one or more characteristics of individuals on the basis of knowledge about related characteristics. For example, common-sense observation has taught most of us that the amount of time we practice learning something is somewhat predictive of how well we perform on that thing we are trying to master. Our bowling score tends to improve (up to a point) in relationship to the amount of time we spend practicing bowling. In the social sciences however, we are often interested in predicting less obvious outcomes. For example, we may be interested in predicting how much a person might be expected to use a computer on the basis of a possible relationship between computer usage and other characteristics such as anxiety in using machines, mathematics aptitude, spatial visualization skills, etc. Often we have not even observed the relationships but instead must simply hypothesize that a relationship exists. In addition, we must hypothesize or assume the type of relationship between our variables of interest. Is the relationship a linear one? Is it a curvilinear one?

Multiple regression analysis is a method for examining the relationship between one continuous variable of interest (the dependent or criterion variable) and one or more independent (predictor) variables. Typically we assume a linear relationship of the type

$$Y_i = B_1X_{i1} + B_2X_{i2} + \ldots + B_kX_{ik} + B_0 + E_i \qquad (4.1)$$

where

Y_i is the score obtained for individual i on the dependent variable,
$X_{i1} \ldots X_{ik}$ are scores obtained on k independent variables,
$B_1 \ldots B_k$ are weights (regression coefficients) of the k independent variables which maximize the relationship with the Y scores,
B_0 is a constant (intercept) and E_i is the error for individual i.

In the above equation, the error score E_i reflects the difference between the subject's actual score Yi and the score which is predicted on the basis of the weighted combination of the independent variables. That is,

$$Y'_i - Y_i = E_i. \qquad (4.2)$$

where Y'_i is predicted from

$$Y'i = B_1X_{i1} + B_2X_{i2} + \ldots + B_kX_{ik} + B_0 \qquad (4.3)$$

In addition to assuming the above general linear model relating the Y scores to the X scores, we usually assume that the E_i scores are normally distributed.

When we complete a multiple regression analysis, we typically draw a sample from a population of subjects and observe the Y and X scores for the subjects of that sample. We use that sample data to estimate the weights (B's) that will permit us the "best" prediction of Y scores for other individuals in the population for which we only have knowledge of their X scores. For example, assume we are interested in predicting the scores that individuals make on a paper and pencil final examination test in a statistics course in graduate college. We might hypothesize that students who, in the past, have achieved higher grade point averages as undergraduates would likely do better on a statistics test. We might also suspect that students with higher mathematics aptitudes as measured by the mathematics score on the Graduate Record Examination would do better than students with lower scores. If undergraduate GPA and GRE-Math combined are highly related to achievement on a graduate statistics grade, we could use those two variables as predictors of success on the statistics test. Note that in this example, the GRE and undergraduate GPA are obtained for individuals quite some time before they even enroll in the statistics course! To find that weighted combination of GRE and GPA scores which "best" predicts the graduate statistics grades of students, we must observe the actual grades obtained by a sample of students that take the statistics course.

Notice that in our linear prediction model, we are going to obtain, for each individual, a single predictor score that is a weighted combination of independent variable scores. We could, in other words, write our prediction equation as

$$Y'_i = C_i + B_0 \qquad (4.4)$$

The Linear Regression Equation

where

$$C_i = \sum_{j=1}^{k} B_i X_I \quad (4.5)$$

You may recognize that (4.3) above is a simple linear equation. The product–moment correlation between Yi and Ci in (4.3) is an index of the degree to which the dependent and composite score are linearly related. In a previous chapter we expressed this relationship with r_{xy} and the proportion of variance shared as r^2_{xy}. When x is replace by a weighted composite score C, we differentiate from the simple product–moment correlation by use of a capital r, that is $R_{y.1,2,...,k}$ with the subscripts after the period indicating the k independent variables. The proportion of variance of the Y scores that is predicted by weighted composite of X scores is, similarly, $R^2_{y.1,2,...,k}$.

We previously learned that, for one independent variable, the "best" weight (B) could be obtained from

$$B = r_{xy} S_y / S_x. \quad (4.6)$$

We did not, however, demonstrate exactly what was meant by the best fitting line or best B. We need to learn how to calculate the values of B when there is more than one independent variable and to interpret those weights.

In the situation of one dependent and one independent variable, the regression line is said to be the "best" fitting line when the squared distance of each observed Y score summed across all Y scores is a minimum. The figure on the following page illustrates the "best fitting" line for the pairs of x and y scores observed for five subjects. The line represents, of course, the equation

$$Y'_i = BX_i + B_0 \quad (4.7)$$

That is, the predicted Y value for any value of X. (See Chap. 3 to review how to obtain B and B_0). Since we have defined error (E_i) as the difference between the observe dependent variable score (Y_i) and the predicted score, then our "best fitting" line is drawn such that

$$\sum_{i=1}^{n} E_i^2 = \sum_{i=1}^{n} (Y_i - Y'_i)^2 \text{ is a minimum.} \quad (4.8)$$

We can substitute our definition of Y'_i from (4.7) above in (4.8) above and obtain

$$G = \sum_{i=1}^{n} [Y_i - (BX_i + B_0)]^2 = \text{a minimum} \quad (4.9)$$

Expanding (4.5) yields

$$G = \sum_{i=1}^{n} Y_i^2 + \sum_{i=1}^{n} (BX_i + B_0)^2 - 2 \sum_{i=1}^{n} Y_i(BX_i + B_0)$$

$$= \sum_{i=1}^{n} Y_i^2 + \sum_{i=1}^{n} (B^2 X_i^2 + B_0^2 + 2B_0 B X_i) - 2B \sum_{i=1}^{n} Y_i X_i - 2B_0 \sum_{i=1}^{n} Y_i$$

(4.10)

or

$$G = \sum_{i=1}^{n} Y_i^2 + B^2 \sum_{i=1}^{n} X_i^2 + nB_0^2 + 2B_0 B \sum_{i=1}^{n} X_i - 2B \sum_{i=1}^{n} Y_i X_i - 2B_0 \sum_{i=1}^{n} Y_i$$
$$= \text{a minimum.}$$

(4.11)

Notice that the function G is affected by two unknowns, B_0 and B. There is one pair of these values which makes G a minimum value _ any other pair would cause G (the sum of squared errors) to be larger. But how do we determine B and B_0 that guarantees, for any observed sample of data, a minimum G? To answer this question requires we learn a little bit about minimizing a function. We will introduce some very elementary concepts of Calculus in order to solve for values of B and B_0 that minimize the sum of square errors.

Least Squares Calculus

Definitions:

Definition 1 A function (f) is a correspondence between the objects of one class and those of another which pairs each member of the first class with one and only one member of the second class. We have several ways of specifying functions, for example, we might provide a complete cataloging of all the associated pairs, e.g.

$$\frac{\text{Class 1 (x)} \mid 1\ 2\ 3\ 4\ 5}{\text{class 2 f (x)} \mid 3\ 5\ 7\ 9\ 11}$$

where class 2 values are a function of class 1 values.

Another way of specifying a function is by means of a set of ordered pairs, e.g.

$$\{(1,3),\ (2,5),\ (3,7),\ (4,9),\ (5,11)\}$$

Fig. 4.1 A simple function map

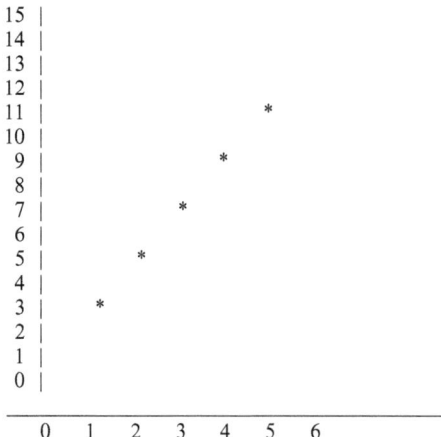

We may also use a map or graph such as (Fig. 4.1)
Finally, we may use a mathematical formula:

$$f(x) = 2X + 1 \text{ where } X = 1, 2, 3, 4, 5$$

Definition 2 Given a specific member of the first class, say X, the member of the second class corresponding to this first class member, designated by f(X), is said to be the value of the function at X.

Definition 3 The set of all objects of the first class is said to be the domain of the function. The set of all objects of the second class is the range of the function f(X).

In our previous example under definition 1, the domain is the set of numbers (1,2,3,4,5) and the range is (3,5,7,9,11). As another example, let X = any real number from 1 to 5 and let f(X) = 2X + 1. Then the domain is

$$\{X : 1 \leq X \leq 5\} \text{ and the range is}$$
$$\{f(X) : 3 \leq f(X) \leq 11\}.$$

Definition 4 The classes of objects or numbers referred to in the previous definitions are sometimes called variables. The first class is called the independent variable and the second class is called the dependent variable.

Definition 5 A quantity which retains a fixed value throughout the course of a discussion is called a constant. Some constants retain the same values in all discussions, c.g.

$$B = c/d = 3.1416\ldots, \text{ and}$$
$$e = \text{limit as } x \, 6 \, 4 \text{ of } (1 + X)^{1/X} = 2.7183\ldots .$$

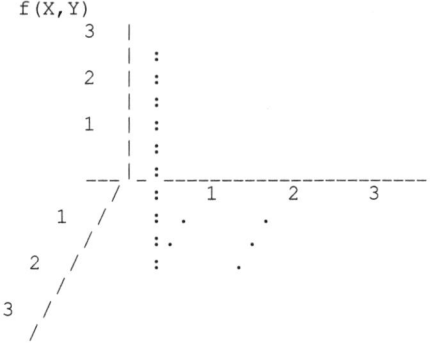

Fig. 4.2 A function map in three dimensions

Other constants retain the save values in a given discussion but may vary from one discussion to another. For example, consider the function

$$f(X) = bX + a. \tag{4.12}$$

In the example under definition 1, $b = 2$ and $a = 1$. If $b = -2$ and $a = 3$ then the function becomes

$$f(x) = -2X + 3.$$

If X is continuous or an infinite set, complete listing of the numbers is impossible but a map or formula may be used. Now consider

$$\frac{X \mid 1\ 2\ 2\ 3}{f(X)\mid 3\ 5\ 7\ 4}$$

This is not a legitimate function as by definition there is not a one and only one correspondence of members.

Sometimes the domain is itself a set of ordered pairs or the sub-set of a plane (Fig. 4.2). For example

The domain of $\{ (X, Y) : 0 \leq X \leq 2 \ \& \ 0 \leq Y \leq 2 \}$

$$f(X, Y) = 2X + Y + 1$$

Range of $\{ 1 \leq f(X, Y) \leq 7 \}$

Finding a Change in Y Given a Change in X for $Y = f(X)$

It is often convenient to use Y or some other letter as an abbreviation for f(X).

Definition 6 ΔX represents the amount of change in the value of X and ΔY represents the corresponding amount of change in the value of $Y = f(X)$. ΔX and

Finding a Change in Y Given a Change in X for Y = f(X)

ΔY are commonly called increments of change or simply increments. For example, consider $Y = f(X) = X^2$ where:

Domain is $\{X : -\infty < X < +\infty\}$
Now let $X = 5$. Then $Y = f(X) = 25$. Now let $\Delta X = +2$. Then $Y = +24$. Or let $\Delta X = -2$ then $Y = -16$. Finally, let $\Delta X = 1/2$ then $Y = 5.25$.

Trying a different starting point $X = 3$ and using the same values of X we would get:

if $X = 3$
and $\Delta X = +2$ then $Y = +16$
$\Delta X = -2$ then $Y = -8$
$\Delta X = .5$ then $Y = 3.25$

It is impractical to determine the increment in Y for an increment in X in the above manner for the general function $Y = f(X) = X^2$. A more general solution for Y is obtained by writing

$$Y + \Delta Y = f(X + \Delta X) = (X + \Delta X)^2$$

or, solving for Y by subtracting Y from both sides gives

$$Y = (X + \Delta X)^2 - Y \qquad (4.13)$$

or $Y = X^2 + \Delta X^2 + 2X\ \Delta X - Y$
or $Y = X^2 + \Delta X^2 + 2X\ \Delta X - X^2$
or $Y = 2X\ \Delta X + \Delta X^2$

Now using this formula:
If $X = 5$ and $\Delta X = 2$ then $Y = +24$ or if $X = 5$ and $\Delta X = -2$ then $Y = -16$. These values are the same as we found by our previous calculations!

Relative Change in Y for a Change in X

We may express the relative change in a function with respect to a change in X as the ratio

$$\frac{\Delta Y}{\Delta X}$$

For the function $Y = f(X) = X^2$ we found that $Y = 2X\ \Delta X + \Delta X^2$
Dividing both sides by ΔX we then obtain

$$\frac{\Delta Y}{\Delta X} = 2X + \Delta X \qquad (4.14)$$

For example, when $X = 5$ and $X = +2$, the relative change is

$$\frac{\Delta Y}{\Delta X} = \frac{24}{2} = 2(5) + 2 = 12$$

The Concept of a Derivative

We may ask what is the limiting value of the above ratio ($\Delta Y/\Delta X$) of relative change is as the increment in X (ΔX) approaches 0 ($\Delta X \to 0$). We use the symbol

$$\frac{dY}{dX} \text{ to represent this limit.}$$

We note that for the function $Y = X^2$, the relative change was

$$\frac{\Delta Y}{\Delta X} = 2X + \Delta X. \tag{4.15}$$

If ΔX approaches 0 then the limit is

$$\frac{dY}{dX} = 2X.$$

Definition 7 The derivative of a function is the limit of a ratio of the increment of change of the function to the increment of the independent variable when the latter increment approaches 0 as a limit. Symbolically,

$$\frac{dY}{dX} = \lim_{\Delta X \to 0} \frac{\Delta Y}{\Delta X} = \lim_{\Delta X \to 0} \frac{f(X + \Delta X) - f(X)}{\Delta X}$$

Since $Y + \Delta Y = f(X + \Delta X)$ and $Y = f(X)$ then $\Delta Y = f(X + \Delta X) - f(X)$ and the ratio

$$\frac{\Delta Y}{\Delta X} = \frac{f(X + \Delta X) - f(X)}{\Delta X} \tag{4.16}$$

Example: $\quad Y = X^2 \, dY/dX = ?$

Finding a Change in Y Given a Change in X for Y = f(X)

$$dY/dX = \lim_{\Delta X \to 0} \frac{f(X + \Delta X) - f(X)}{\Delta X}$$

$$= \lim_{\Delta X \to 0} \frac{X^2 + \Delta X^2 + 2X\,\Delta X - X^2}{\Delta X}$$

$$= \lim_{\Delta X \to 0} \Delta X + 2X$$

or $\dfrac{dY}{dX} = 2X$

Some Rules for Differentiating Polynomials

Rule 1 *If* $Y = CX^n$, *where n is an integer, then*

$$\frac{dY}{dX} = nCX^{n-1} \qquad (4.17)$$

For example, let $C = 7$ *and* $n = 4$ *then* $Y = 7X^4$.

$$\frac{dY}{dX} = (4)(7)X^3$$

Proof

$$\frac{dY}{dX} = \lim_{\Delta X \to 0} \frac{C(X + \Delta X)^n - CX^n}{\Delta X}$$

since $(a + b)^n = \sum_{r=0}^{n} a^r b^{n-r}$

then

$$\frac{dY}{dX} = \lim_{\Delta X \to 0} \left[C\left(\sum_{n}^{n} X^n \Delta X^{n-n}\right) + C\left(\sum_{n-1}^{n} X^{n-1} \Delta X^1\right) \right.$$

$$+ C\left(\sum_{n-2}^{b} X^{n-2} \Delta X^2\right) + \ldots + C\left(\sum_{0}^{b} X^0 \Delta X^n\right)$$

$$\left. - CX^n \right] / \Delta X$$

$$= \lim_{\Delta X \to 0} \left[CX^n + C^n X^{n-1} \Delta X + C\frac{n(n\_1)}{2} X^n {}^2\Delta X^2 \right.$$

$$\left. + \ldots + C\,\Delta X^n - CX^n \right] / \Delta X$$

$$= \lim_{\Delta X \to 0} CnX^{n-1} + \frac{n(n\_1)}{2} X^{n-2} \Delta X + \ldots + C\,\Delta X^{n-1}$$

or

$$\frac{dY}{dX} = CnX^{n-1} \quad \text{(End of Proof)}$$

Rule 1.a If $Y = CX$ then $dY/dX = C$ (4.18)

since by Rule 1 $dY/dX = (1)CX^0 = C$

Rule 1.b *If $Y = C$ then $dY/dX = 0$*
Note that dY/dX of CX^0 is $(0)CX^{-1} = 0$.

Rule 2 *If $Y = U + V - W$ where U, V and W are functions of X, then:*

$$\frac{dY}{dX} = \frac{dU}{dX} + \frac{dV}{dX} - \frac{dW}{dX} \quad (4.19)$$

Example : Consider $Y = 4X^2 - 4X + 1$
  Let $U = f(X) = 4X^2$ and
  $V = f(X) = -4X$ and
  $W = f(X) = 1$.
  Applying Rules 1 and 2 we have

$$\frac{dY}{dX} = 8X - 4$$

Rule 3 *If $U = V^n$ where V is a function of X then*

$$\frac{dY}{dX} = nV^{n-1}\frac{dU}{dX} \quad (4.20)$$

Example: Consider $Y = (2X - 1)^2$
  Let $V = (2X - 1)$ and $n = 2$
  Then

$$\frac{dY}{dX} = 2(2X - 1)(2) = 8X - 4$$

Another Example. Let $Y = \sum_{i=1}^{N}(3X + W_i)^2$

where W_i and N are variable constants, that is, in one discussion $N_1 = 3$ and $W_1 = 2$ or $W_2 = 4$ and $W_3 = 3$.

If, for example, $X = 0$, $Y = 2^2 + 4^2 + 3^2 = 29$
or, if $X = 1$ then $Y = 5^2 + 7^2 + 6^2 = 110$
Now we ask, $dY/dX = ?$

Solution:

$$\frac{dY}{dX} = \sum_{i=1}^{N} 2(3X + W_i)(3)$$

because $Y = (3X + W_1)^2 + (3X + W_2)^2 + (3X + W_3)^2$
and applying Rules 2 and 3 we get:

$$\frac{dY}{dX} = 6 \sum_{i=1}^{N} (3X + W_i)$$

$$= 6 \sum_{i=1}^{N} 3X + 6 \sum_{i=1}^{N} W_i$$

$$= 6[N(3X)] + 6 \sum_{i=1}^{N} W_i$$

or

$$\frac{dY}{dX} = 18NX + 6 \sum_{i=1}^{N} W_i$$

Geometric Interpretation of a Derivative

The figure below presents a graphical representation of a function $Y = f(X)$ (the curved line). Two points on the function are denoted as $P(X,Y)$ and $P(X + X, Y + Y)$. A straight line, a secant line, is drawn through the two points. Notice that if X becomes smaller (and therefore the corresponding Y becomes smaller) that the secant line approaches a tangent line at the point $P(X,Y)$. We review:

$$f(X) = Y$$

$$f(X + \Delta X) = (Y + \Delta Y) \quad \text{or} \quad f(X + \Delta X) - f(X) = Y$$

and

$$\frac{f(X + \Delta X) - f(X)}{\Delta X} = \frac{\Delta Y}{\Delta X}$$

Note that ΔY/ΔX give rise over run or the slope of the of the secant line through two points on the function. Now if X → 0, then P' approaches P and the secant line approaches a tangent at the point P. Therefore the dY/dX is the slope of the tangent at P or X.

We will now use the derivative in determining maximum points on a function.

Finding the Value of X for Which f(X) Is Least

Given the function $f(X) = Y = X^2 - 3X$ where $-\infty < X < +\infty$ we may present the function as in Fig. 4.3 below.

For the function, we may obtain some values of Y corresponding to a selected set of X values:

$$\begin{array}{c|c} X & -2 \;-1 \;\;0+1+2+3+4+5 \\ \hline Y & 10 \;\;\;4 \;\;\;0-2-2 \;\;0+4+10 \end{array}$$

Then the derivative

$$\frac{dY}{dX} = 2X - 3 \text{ which is the slope of the tangent at any point X.}$$

Setting the slope (dY/dX) equal to zero we obtain the minimum value of X, that is,

$0 = 2X - 3$ and therefore $X = 1.5$ for a minimum Y value.

Another Example of a Minimum
Given a collection of score values X

$$\{X \mid 16, 8, 10, 4, 12\}$$

we ask for what value of A is f(A) a minimum if

$$f(A) = \sum_{i=1}^{5} (Xi - A)^2 \;?$$

First, examine the f(A) for various values of A, for example:

if $A = 5$ then $f(A) = 11^2 + 3^2 + 5^2 + (-1)^2 + 72$
if $A = 7$ then $f(A) = 9^2 + 1^2 + 3^2 + (-3)^2 + 52$
if $A = 8$ then $f(A) = 8^2 + 0^2 + 2^2 + (-4)^2 + 42$
etc.

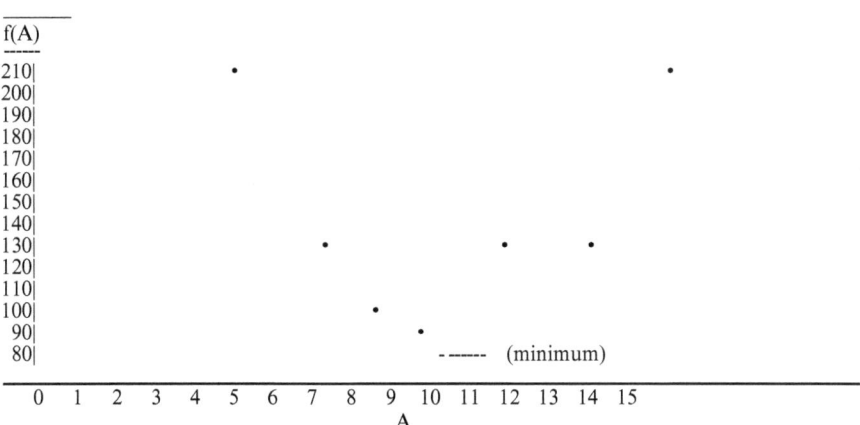

Fig. 4.3 The minimum of a function derivative

A plot of the function f(A) is presented below for the values (Fig. 4.3)

$$\begin{array}{c|ccccccc} A & 5 & 7 & 8 & 9 & 11 & 13 & 15 \\ \hline f(A) & 205 & 125 & 100 & 85 & 85 & 125 & 205 \end{array}$$

The derivative of the f(A) with respect to A is

$$\frac{d\,f(A)}{dA} = \sum_{i=1}^{5} 2(X_i - A)(-1) \qquad (4.21)$$

and to obtain the minimum slope point we obtain

$$0 = \sum_{i=1}^{5} -2(X_i - A) = \sum_{i=1}^{5} X_i - 5A \qquad (4.22)$$

or

$$A = \sum_{i=1}^{5} X_i / 5$$

Therefore $A = (16 + 8 + 10 + 4 + 12)/5 = 10$
and $f(A) = 80$ is a minimum.

A Generalization of the Last Example

We will use derivation to prove that given any collection of X values $X1, X2, \ldots, Xi, \ldots, XN$ that

$Y = \sum_{i=1}^{N}(X_i - A)^2$ is least when $A = X$

As before, the derivative of Y with respect to A is

$$\frac{dY}{dA} = \sum_{i=1}^{N} 2(X_i - A)(-1) = -2\sum_{i=1}^{N}(X_i) - 2NA(-1)$$

Therefore if we set the derivative to zero we obtain

$$0 = -2\sum_{i=1}^{N} X_i + 2NA$$

or

$$0 = -\sum_{i=1}^{N} X_i + NA$$

then

$$A = \sum_{i=1}^{N} X_i / N \qquad (4.23)$$

which by definition is $\overline{X}$.

Partial Derivatives

Given a function in two independent variables:

$$Y = f(X, Z)$$

we may create a graph as shown in Fig. 4.2 above. Y, the function, is shown as the vertical axis and X and Z are shown as horizontal axis. Note the line in the figure which represents the map of f(X,Z) when one considers only one value of Z.

When we study functions of this type with one variable treated as a constant, the derivative of the function is called a partial derivative.

Suppose the function has a minimum and that it occurs at $X = A$ and $Z = B$, that is, f(A,B) is a minimum value of Y. We may obtain the derivative of $Y = f(A,Z)$, that is, treat Z as a constant. This would be the partial derivative $\delta Y/\delta Z$ and may be set equal to 0 to get the minimum at B. Of course, we don't know A. Likewise, $Y = f(X,B)$ and $\delta Y/\delta X$ set equal to 0 will give A. Here we don't know B.

We can however, by simultaneous equations, where A and B are set to 0, find a minimum of X and Z to give the Y minimum.

For example, let $Y = f(X, Z) = X^2 + XZ + Z^2 - 6X + 2$.

Then

$$\frac{\delta Y}{\delta X} = 2X + Z - 6 = 0 \qquad (4.24)$$

and

$$\frac{\delta Y}{\delta Z} = X + 2Z = 0 \qquad (4.25)$$

or $X = -2Z$ for equation (4.25) and substituting in (4.24) gives

$$-4Z + Z = 6 \text{ or } Z = -2$$

and therefore $X = +4$. These values of Z and X are the values of A and B to produce a minimum for $Y = f(A,B)$.

Least Squares Regression for Two or More Independent Variables

In this section we want to use the concepts of partial derivation to obtain solutions to the B values in

$$Y'_i = B_1 X_{i,1} + B_2 X_{i,2} + B_0 \qquad (4.26)$$

such that the sum of $(Y - Y')^2$ is a minimum.

As an example, assume we have a situation in which values of Y_i represent Grade Point Average (GPA) score of subject (i) in his or her freshman year at college. Assume that the $X_{i,1}$ score is the high school GPA and that the $X_{i,2}$ is an aptitude test score. Our population of subjects may be "decomposed" into sub-populations of Y scores that correspond to given values of X_1 and X_2. Figure 4.4 depicts the distributions of Y scores for combinations of X_1 and X_2. We will assume:

1. The experience pool of the available data is a random sample of $(Y, X_1$ and $X_2)$ triplets from a universe of such triplets,
2. The universe is capable of decomposition into sub-universes of triplets have like X_1 and X_2 values but differing in Y values,
3. The Y means for the sub-universes fall on a plane, that is,

$$\mu_{Y,12} = \beta_1 X_1 + \beta_2 X_2 + \beta_0 \qquad (4.27)$$

Now we use the data to estimate β_1, β_2 and β_0 by finding those values of B_1 and B_2 and B_0 in:

$$Y' = B_1 X_1 + B_2 X_2 + B_0 \qquad (4.28)$$

Fig. 4.4 Three Dimension View of GPA (RED), High School GPA (Green) and Aptitude (Black)

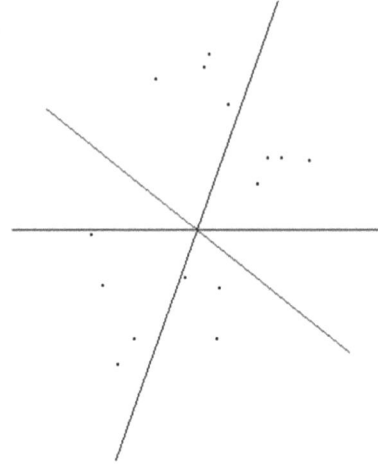

which minimize

$$G = \sum_{i=1}^{N} (Y_i - Y'_i)^2$$

or
$$G = \sum_{i=1}^{N} [Y_i - (B_1 X_{i,1} + B_2 X_{i,2} + B_0)]^2 \quad (4.29)$$

The steps to our solution are:

1. Find the partial derivatives and equate them to 0.

$$\frac{\delta G}{\delta B_1} = 2 \sum_{i=1}^{N} [Y_i - (B_1 X_{i,1} + B_2 X_{i,2} + B_0)](-X_{i,1})$$

$$\frac{\delta G}{\delta B_2} = 2 \sum_{i=1}^{N} [Y_i - (B_1 X_{i,1} + B_2 X_{i,2} + B_0)](-X_{i,2})$$

$$\frac{\delta G}{\delta B_0} = 2 \sum_{i=1}^{N} [Y_i - (B_1 X_{i,1} + B_2 X_{i,2} + B_0)](-1)$$

Now equating to 0 and simplifying results in the following three "normal" equations:

$$\sum_{i=1}^{N} Y_i X_{i,1} = B_1 \sum_{i=1}^{N} X_{i,1}^2 + B_2 \sum_{i=1}^{N} X_{i,1} X_{i,2} + B_0 \sum_{i=1}^{N} X_{i,1} \quad (4.30)$$

$$\sum_{i=1}^{N} Y_i X_{i,2} = B_1 \sum_{i=1}^{N} X_{i,1} X_{i,2} + B_2 \sum_{i=1}^{N} X_{i,2}^2 + B_0 \sum_{i=1}^{N} X_{i,2} \quad (4.31)$$

$$\sum_{i=1}^{N} Y_i = B_1 \sum_{i=1}^{N} X_{i,1} + B_2 \sum_{i=1}^{N} X_{i,2} + NB_0 \quad (4.32)$$

2. Use the data to obtain the various sums, sums of squared values, and sums of products needed. Substitute them in the above equations (4.30), (4.31) and (4.32) and solve the equations simultaneously for B_1, B_2 and B_0.
3. Substitute obtained values of B_1, B_2 and B_0 into equation (4.29) to get the regression equation.
4. If an index of accuracy of prediction is desired, calculate

$$\sum_{i=1}^{N} y'^2_i \text{ and obtain } R^2_{y.12} = \frac{\sum_{i=1}^{N} y'^2_i}{\sum_{i=1}^{N} y^2_i} \quad (4.33)$$

where the y'_i and y_i scores are deviations from the mean Y value.

Matrix Form for Normal Equations Using Raw Scores

Equations (4.4), (4.5) and (4.6) above may be written more conveniently in matrix form as:

$$\left[\sum_{i=1}^{N} Y_i X_{i,1} \quad \sum_{i=1}^{N} Y_i X_{i,2} \quad \sum_{i=1}^{N} Y_i \right] =$$

$$[B_1 B_2 B_0] \begin{vmatrix} \sum_{i=1}^{N} X_{i,1}^2 & \sum_{i=1}^{N} X_{i,1} X_{i,2} & \sum_{i=1}^{N} X_{i,1} \\ \sum_{i=1}^{N} X_{i,1} X_{i,2} & \sum_{i=1}^{N} X_{i,2}^2 & \sum_{i=1}^{N} X_{i,2} \\ \sum_{i=1}^{N} X_{i,1} & \sum_{i=1}^{N} X_{i,2} & N \end{vmatrix}$$

or $[Y'X]_{1 \times (K+1)} = [B]'_{1 \times (K+1)} [X'X]_{(K+1)(K+1)}$
and leaving off the sizes of the matrices gives simply

$$[Y'X] = [B]' [X'X].$$

If we post-multiply both sides of this equation by $[X'X]^{-1}$ we obtain

$$[Y'X][X'X]-1 = [B]' \quad (4.34)$$

We note that B_0 may also be obtained from

$$B_0 = \overline{Y} - (B_1\overline{X}_1 + \ldots + B_k\overline{X}_k) \quad (4.35)$$

or in matrix notation

$$B_0 = \overline{Y} - [B]'[\overline{X}] \quad (4.36)$$

where $[\overline{X}] = (1/N)[X]$

Matrix Form for Normal Equations Using Deviation Scores

The prediction (regression) equation above may be written in deviation score form as

$$y' = B_1 x_{i,1} + B_2 x_{i,2} \quad (4.37)$$

and solve for $G = \sum_{i=1}^{N}(y_i - y'_i)^2$ as a minimum. In deviation score form there is no B_0 since the means of deviation scores are always 0.

The partial derivatives of G with respect to B_1 and B_2 may be written as follows: with $\frac{\delta G}{\delta B_1} = 0$ and $\frac{\delta G}{\delta B_2} = 0$ we obtain

$$B_1 \sum_{i=1}^{N} x_{i,1}^2 + B_2 \sum_{i=1}^{N} x_{i,1} x_{i,2} = \sum_{i=1}^{N} y_i x_{i,1}$$

and

$$B_1 \sum_{i=1}^{N} x_{i,1} x_{i,2} + B_2 \sum_{i=1}^{N} x_{i,2}^2 = \sum_{i=1}^{N} y_i x_{i,2}$$

or in matrix notation

$$[B_1 B_2] \begin{vmatrix} \sum_{i=1}^{N} x_{i,1}^2 & \sum_{i=1}^{N} X_{i,1} X_{i,2} \\ \sum_{i=1}^{N} X_{i,1} X_{i,2} & \sum_{i=1}^{N} X_{i,2}^2 \end{vmatrix}$$

$$= \left[\sum_{i=1}^{N} y_i X_{i,1} \quad \sum_{i=1}^{N} y_i X_{i,2} \right]$$

or simply

$$[B]' [x'x] = [y'x]'$$

and

$$[B]' = [y'x]' [x'x]^{-1} \qquad (4.38)$$

Matrix Form for Normal Equations Using Standardized Scores

The regression equation from above may be written in terms of standardized (z) scores as

$$z'_y = \beta_1 z_1 + \beta_2 z_2 \qquad (4.39)$$

The function to be minimized is $G = \sum_{i=1}^{N} \left(z_y - z'_y\right)^2$.

We obtain the partial derivatives of G with respect to β_1 and β_2 as before and set them to zero. The equations obtained are then

$$\beta_1 \sum_{i=1}^{N} z^2_1 + \beta_2 \sum_{i=1}^{N} z_1 z_2 = \sum_{i=1}^{N} z_y z_1$$

and

$$\beta_1 \sum_{i=1}^{N} z_1 z_2 + \beta_2 \sum_{i=1}^{N} z^2_2 = \sum_{i=1}^{N} z_y z_2$$

If we divide both sides of the above equations by N we obtain

$$\beta_1 + \beta_2 r_{1,2} = r_{y,1}$$

$$\beta_1 r_{1,2} + \beta_2 = r_{y,2}$$

or

$$[\beta_1 \ \beta_2] \begin{vmatrix} 1 & r_{1,2} \\ r_{1,2} & 1 \end{vmatrix} = [r_{y,1} \ r_{y,2}]$$

or more simply as

$$[\beta]' \, [r_{xx}] = [r_{y,x}]'$$

and therefore

$$[\beta]' = [r_{y,x}]' \, [r_{x,x}]^{-1} \qquad (4.40)$$

Equations in the previous discussion are general forms for solving the regression coefficients $B_1,\ldots,B_{k+1}$ in raw score form, the $B_1,\ldots,B_k$ coefficients in deviation score form or the $\beta_1,\ldots,\beta_k$ coefficients in standardized score form. In each case, the B's or Betas are obtained by multiplication of an inverse matrix times the vector of cross-products or correlations. When there are more than two independent variables, the inverse of the matrix becomes laborious to obtain by hand. Computers are generally available however, which makes the chore of obtaining an inverse much easier.

You should remember that the independent variables must, in fact, be independent. That is, one independent variable cannot be a sum of one or more of the other independent variables. If the assumption of independence is violated, the inverse of the matrix may not exist! In some cases, although the variables are independent, they may nevertheless correlate quite highly among themselves. In such cases (high colinearity among independent variables), the computation of the inverse matrix may be difficult and result in considerable error. If the determinant of the matrix is very close to zero, your results should be held suspect!

We will see in latter sections that the inverse of the matrix of independent variable cross-products, deviation cross-products or correlations may be used to estimate the standard errors of regression coefficients and the covariances among the regression coefficients.

Hypothesis Testing in Multiple Regression

Testing the Significance of the Multiple Regression Coefficient

The multiple regression coefficient $R_{Y,12\ldots k}$ is an index of the degree to which the dependent and weighted composite of the independent variables correlate. The square of the coefficient indicates the proportion of variance of the dependent variable which is predicted by the independent variables. The $R^2_{Y,12\ldots k}$ may be obtained from

$$R^2_{Y,1..k} = [\beta]' \, [r_{y,x}] \text{ that is}$$

$$R^2 = \beta_1 r_{y,1} + \beta_2 r_{y,2} + \ldots + \beta_k r_{y,k} \tag{4.41}$$

Since R^2 is a sample statistic which estimates a population parameter, it may be expected to vary from sample to sample and has a standard error.

The total sum of squares of the dependent variable Y may be partitioned into two main sources of variability:

1. The sum of squares due to regression with the independent variables (SS_{reg}) and
2. The sum of squares due to error or unexplained variance (SS_e).

We may estimate these values by

(a) $SS_{reg} = SS_Y R^2_{Y.12\ldots k}$ and
(b) $SS_e = SS_Y (1 - R^2_{Y.12\ldots k})$

Associated with each of these sums of squares are degrees of freedom. For the SS_{reg} the degrees of freedom is the number of independent variables, K. For the SS_e the degrees of freedom are $N - K - 1$, that is, the degrees of freedom for the variance of Y minus the degrees of freedom for regression. Since the sum of squares for regression and error are independent, we may form an F-ratio statistic as

$$F = \frac{MS_{reg}}{MS_e} = \frac{Ss_{reg}/K}{Ss_e/(N-K-1)} = \frac{R^2_{Y,1..k}}{\left(1 - R^2_{Y,1..k}\right)} \cdot \frac{N-K-1}{K} \tag{4.42}$$

The probability of the F statistic for K and (N-K-1) degrees of freedom may be estimated or values for the tails obtained from tables of the F distribution. If the probability of obtaining an F statistic as large or larger than that calculated is less than the alpha level selected, the hypothesis that $R^2 = 0$ in the population may be rejected.

The Standard Error of Estimate

The following figure illustrates that for every combination of the independent variables, there is a distribution of Y scores. Since our prediction equation based on a sample of observations yields only a single Y value for each combination of the independent variables, there are obviously some predicted Y scores that are in error. We may estimate the variability of the Y scores at any combination of the X scores. The standard deviation of these scores for a given combination of X scores is called the Standard Error of Estimate. It is obtained as

$$S_{Y.X} = (SS_e/(N-K-1))^{1/2} \tag{4.43}$$

Testing the Regression Coefficients

Just as we may test the hypothesis that the overall multiple regression coefficient does not depart significantly from zero, so may we test the hypothesis that a regression coefficient B does not depart significantly from zero. Note that if we conclude that the coefficient does not depart from zero, we are concluding that the associated variable for that coefficient does not contribute significantly to explaining (predicting) the variance of Y.

The regression coefficients have been expressed both in raw score form (B's) and in standardized score form (β's). We may convert from one form to the other using

$$B_j = \beta_j S_Y / S_j \quad (4.44)$$

or

$$\beta_j = B_j S_j / S_Y$$

Since these coefficients are sample statistics, they have a standard error. The standard error of a regression coefficient may be obtained as the square root of:

$$S^2_{B_j} = \frac{S^2_{Y.X}}{SS_{X_j}(1 - R^2_{j,1..(k-1)})} \quad (4.45)$$

where $S^2_{Y.X}$ is the standard error of estimate and SS_X is the sum of squares for the jth variable,

$R^2_{j,1..(k-1)}$ is the squared multiple correlation of the jth independent variable regressed on the $K - 1$ remaining independent variables.

In using the above method to obtain the standard errors of regression coefficients, it is necessary to obtain the multiple correlation of each independent variable with the remaining independent variables.

Another method of obtaining the standard errors of B's is through use of the inverse of the matrix of deviation score cross-products among the independent variables. We indicated this matrix as

$$[x'x]^{-1}$$

in our previous discussion. If we multiply this matrix by the variance of our error of estimate $S^2_{Y.X}$ the resulting matrix is the variance-covariance matrix of regression coefficients. That is

$$[C] = S^2_{Y.X}[x'x]^{-1} \quad (4.46)$$

The diagonal elements of [C], that is, $C_{1,1}, C_{2,2},...,C_{k,k}$ are the variances of the B regression coefficients and the off-diagonal values are the covariances of the regression coefficients for independent variables.

Hypothesis Testing in Multiple Regression

To test whether or not the B_j regression coefficient departs significantly from zero, we may use either the t-test statistic or the F-test statistic. The t-test is

$$t = \frac{B_j}{\sqrt{C_{j,j}}} = \frac{B_j}{S_{B_j}} \text{ with N} - \text{K} - 1 \text{ degrees freedom.}$$

Since the t^2 is equivalent to the F test with one degree of freedom in the numerator, we can similarly use the F statistic with 1 and N-K-1 degrees of freedom where

$$F = \frac{B_j^2}{C_{j,j}} \tag{4.47}$$

A third method for examining the effect of a single independent variable is to ask whether or not the inclusion of the variable in the regression model contributes significantly to the increase in the SS_{reg} over the regression model in which the variable is absent. Since the proportion of variance of Y that is accounted for by regression is R^2, we can obtain the proportion of variance accounted for by a variable by

$$R^2_{Y,1..j..K} - R^2_{Y,1..(K-1)} \tag{4.48}$$

The first R^2 equation (we will call it the FULL Model) contains all independent variables. The second (which we will call the restricted model) is the proportion of Y score variance predicted by all independent variables except the jth variable. The difference then is the proportion of variance attributable to the jth variable. The sum of squares of Y for the jth variable is therefore

$$SS_j = SS_Y \left(R^2_{Y,1..j..K} - R^2_{Y,1..(K-1)} \right) \tag{4.49}$$

The mean square for this source of variability is the same as the SS since there is only one degree of freedom. The ratio of the MS_j to MS_e forms an F statistic with 1 and N-K-1 degrees of freedom. That is

$$\begin{aligned} F = \frac{MS_j}{MS_e} &= \frac{SS_Y \left(R^2_{full} - R^2_{restricted} \right)/1}{SS_Y \left(1 - R^2_{full} \right)/(N-K-1)} \\ &= \frac{R^2_{full} - R^2_{restricted}}{1 - R^2_{full}} \cdot \frac{N-K-1}{1} \end{aligned} \tag{4.50}$$

If the independent variable j does not contribute significantly (incrementally) to the variance of Y, the F statistic above will not be significant at the alpha decision level value.

Testing the Difference Between Regression Coefficients

Two variables may differ in the cost of collection. For example, an aptitude test may cost the student or institution more than obtaining a high school grade point average. In selecting one or the other independent variable to use in a regression model, there arises the question as to whether or not two regression coefficients differ significantly between themselves. Since the regression coefficients are sample statistics, the difference between two coefficients B_j and B_k is itself a sample statistic. The regression coefficients B are not independent of one another unless the independent variables themselves are uncorrelated. The standard error of the difference between two coefficients must therefore take into account not only the variance of each coefficient but also their covariance. The variance of differences between two regression coefficients may be obtained as

$$S^2_{B_j - B_k} = C_{j,j} + C_{k,k} - C_{j,k} \tag{4.51}$$

where $C_{j,j}$, $C_{k,k}$ and $C_{j,k}$ are elements of the [C] matrix.

The test for significance of difference between two regression coefficients is therefore

$$t_{(N-K-1)} = \frac{B_j - B_k}{\sqrt{[C_{j,j} + C_{k,k} - C_{j,k}]}} \tag{4.52}$$

Chapter 5
Analysis of Variance

Theory of Analysis of Variance

While the "Student" t-test provides a powerful method for comparing sample means for testing differences between population means, when more than two groups are to be compared, the probability of finding at least one comparison significant by chance sampling error becomes greater than the alpha level (rate of Type I error) set by the investigator. Another method, the method of Analysis of Variance, provides a means of testing differences among more than two groups yet retain the overall probability level of alpha selected by the researcher. Your OpenStat4 package contains a variety of analysis of variance procedures to handle various research designs encountered in evaluation research. These various research designs require different assumptions by the researcher in order for the statistical tests to be justified. Fundamental to nearly all research designs is the assumption that random sampling errors produce normally distributed score distributions and that experimental effects result in changes to the mean, not the variance or shape of score distributions. A second common assumption to most designs using ANOVA is that the sub-populations sampled have equal score variances – this is the assumption of homogeneity of variance. A third common assumption is that the populations sampled have been randomly sampled and are very large (infinite) in size. A fourth assumption of some research designs where individual subjects or units of observation are repeatedly measured is that the correlation among these repeated measures is the same for populations sampled – this is called the assumption of homogeneity of covariance.

When we say we are "analyzing" variance we are essentially talking about explaining the variability of our values around the grand mean of all values. This "Total Sum of Squares" is just the numerator of our formula for variance. When the values have been grouped, for example into experimental and control groups, then each group also has a group mean. We can also calculate the variance of the scores within each of these groups. The variability of these group means around the grand mean of all values is one source of variability. The variability of the scores within the groups is another source of variability. The ratio of the variability of group means to the variability of within-group values is an indicator of how much our total variance is due to differences among our groups. Symbolically, we have "partitioned" our total variability into two parts: variability among the groups and variability within the groups. We sometimes write this as

$$SS_T = SS_B + SS_W \qquad (5.1)$$

That is, the total sum of squares equals the sum of squares between groups plus the sum of squares within groups. The ratio of the SSB to the SSW gives the F statistic. Later we will examine how we might also analyze the variability of scores using a linear equation.

> Once upon a time, a psychologist conducted a survey and gathered considerable amounts of data. However, as is the case many times, the data sat on the shelf gathering dust. But, one year, the psychologist decided to resurrect the data. Not being exactly sure of what to do though, the data was given to a few students to play with and summarize.
>
> Well, as you might expect, one student did it one way, another student did it another way, and a third student even did it entirely different from the other two. Because of this, the psychologist suddenly became interested in a different question and .. proclaimed to the world: "How goes this VARIANCE OF ANALYSIS?"

The Completely Randomized Design

> Why did the statistician do such a horrid job of laying tile on his bathroom floor? He incorrectly PARTITIONED SOME OF THE SQUARES!!

Introduction

Educational research often involves the hypothesis that means of scores obtained in two or more groups of subjects do not differ beyond that which might be expected due to random sampling variation. The scores obtained on the subjects are usually some measure representing relative amounts of some attribute on a dependent variable. The groups may represent different "treatment" levels to which subjects have been randomly assigned or they may represent random samples from some sub-populations of subjects that differ on some other attribute of interest. This treatment or attribute is usually denoted as the independent variable.

Introduction

A Graphic Representation

To assist in understanding the research design that examines the effects of one independent variable (Factor A) on a dependent variable, the following representation is utilized:

		TREATMENT GROUP		
1	2	3	4	5
Y_{11}	Y_{12}	Y_{13}	Y_{14}	Y_{15}
Y_{21}	Y_{22}	Y_{23}	Y_{24}	Y_{25}
.	.	.	.	.
.	.	.	.	.
Y_{n1}	Y_{n2}	Y_{n3}	Y_{n4}	Y_{n5}

In the above figure, each Y score represents the value of the dependent variable obtained for subjects 1, 2,...,n in groups 1, 2, 3, 4, and 5.

Null Hypothesis of the Design

When the researcher utilizes the above design in his or her study, the typical null hypothesis may be stated verbally as "the population means of all groups are equal". Symbolically, this is also written as

$$H_0 : \mu_1 = \mu_2 = \ldots = \mu_k \tag{5.2}$$

where k is the number of treatment levels or groups.

Summary of Data Analysis

The completely randomized design (or one-way analysis of variance design) depicted above requires the researcher to collect the dependent variable scores for each of the subjects in the k groups. These data are then typically analyzed by use of a computer program and summarized in a summary table similar to that below:

SOURCE	DF	SS	MS	F
Groups	k-1	$\sum_{j=1}^{k} n_j (\bar{Y}_j - \bar{Y})^2$	SS / k	$\dfrac{MS_g}{MS_e}$
Error	N-k	$\sum_{j=1}^{k} \sum_{i=1}^{n_j} (Y_{ij} - \bar{Y}_j)^2$	SS / (N-k)	
Total	N-1	$\sum_{j=1}^{k} \sum_{i=1}^{n_j} (Y_{ij} - \bar{Y})^2$		

where Yij is the score for subject i in group j,
Yj is the mean of scores in group j,
Y is the overall mean of scores for all subjects,
nj is the number of subjects in group j, and
N is the total number of subjects across all groups.

Model and Assumptions

Use of the above research design assumes the following:

1. Variance of scores in the populations represented by groups 1,2,...,k are equal.
2. Error scores (which are the source of variability within the groups) are normally distributed.
3. Subjects are randomly assigned to treatment groups or randomly selected from sub-populations represented by the groups.

The model employed in the above design is

$$Y_{ij} = \mu + \mu_j + e_{ij} \tag{5.3}$$

where μ is the population mean of all scores, μ j is the effect of being in group j, and eij is the residual (error) for subject i in group j. In this model, it is assumed that the sum of the treatment effects (αj) equals zero.

Fixed and Random Effects

In the previous section we introduced the analysis of variance for a single independent variable. In our discussion we indicated that treatment levels were usually established by the researcher. Those levels of treatment often are selected to represent specific intervals of a measurement on the independent variable, for example, amount of study time, level of drug dosage, time spent on a task, etc. The independent variable in many one-way analyses of variance may also represent classifications of objects or subjects such as political party, gender, grade level, or country of origin. We suggest more caution in interpretation of outcomes using classification variables since, in these cases, random assignment of subjects from a single population is usually impossible.

There is another situation for analysis of variance. That situation is where the researcher randomly selects levels of the independent variable (or works with objects that have random levels of an independent variable). For example, a researcher may wish to examine the effect of "amount of TV viewing" on student achievement. A random sample of students from a population might be drawn and those subjects tested. The subjects would also be asked to report the number of hours on the average that they watch TV during a week's time. If the analysis of variance is used, the variable "TV time" would be a random variable – the investigator has not assigned hour levels. If the experiment is repeated, the next sample of subjects would most likely represent different levels of TV time, thus the levels randomly fluctuate from sample to sample. For the one-way analysis of variance with the random effects model, the parameters estimated are the same as in the fixed effects model. For the one-way analysis of variance then, the analysis for the random-effects model is exactly the same as for the fixed-effects model (this will NOT be true for two-way and other higher level designs). An additional assumption of the random effects model is that the treatment effects (α) are normally distributed with mean 0 and variance $\sigma e2$. You may recognize that, if both dependent and independent variables are normally distributed and continuous, that the product-moment correlation may be an alternative method of analyzing data of the random-effects model.

Analysis of Variance: The Two-Way, Fixed-Effects Design

A researcher may be interested in examining the effects of two (or more) independent variables on a given dependent variable at the same time. For example, a teacher may be interested in comparing the effects of three types of instruction, e.g. teacher lecture, small group discussion, and self instruction, on student achievement under two other conditions, e.g. students given a pretest and students not given a pretest. There is a possibility that both of these variables contribute to differences in student achievement. In addition, there is the possibility that method of instruction "interacts with" pre-testing conditions. For example, it might be suspected that use of a pretest with teacher lecture method is better than no pretest with teacher lecture but that such a difference would not be observed for the other two methods of instruction. The multi-factor ANOVA designs have the advantage of being able to examine not only the "main" effects of variables hypothesized to affect the dependent variable but also to be able to examine the interaction effects of those variables on the dependent variable.

The data may be conveniently depicted as a rectangle with the levels of one variable on the horizontal axis and the levels of the second variable on the vertical axis. The intersection of each row and column level is a treatment "cell" consisting of njk subjects receiving that combination of treatments. The table below gives the symbolic representation of scores in the two-way design:

		METHOD OF INSTRUCTION		
		Lecture	Group	Self
		$X_{111} = 5$	$X_{112} = 9$	$X_{113} = 5$
	Pretest	$X_{211} = 6$	$X_{212} = 7$	$X_{213} = 12$
		$X_{311} = 4$	$X_{312} = 6$	$X_{313} = 8$
Pretest Condition		$X_{121} = 10$	$X_{122} = 6$	$X_{123} = 4$
	No Pretest	$X_{221} = 12$	$X_{222} = 8$	$X_{223} = 8$
		$X_{321} = 8$	$X_{322} = 9$	$X_{323} = 5$

Using the above data it is possible to consider three seperate one-way ANOVA analyses:

1. An ANOVA of the three methods of instruction,
2. An ANOVA of the two pretesting conditions, and
3. An ANOVA of the 6 cells (treatment combinations).

The two-way ANOVA procedure yields all three in one analysis and provides greater sensitivity for each since the denominator of the F statistic will have the effects of the other two sources of variance already removed. The Summary table for the two-way ANOVA contains:

Analysis of Variance: The Two-Way, Fixed-Effects Design

Source	D.F.	Sum of Squares	F	Parameters Estimated
Rows	R-1	$\sum_{j=1}^{R} n_{j\cdot}(\bar{X}_{\cdot j\cdot}-\bar{X}_{\cdots})^2$	MS_R/MS_e	$\sigma_e^2+\sigma_\alpha^2$
Columns	C-1	$\sum_{k=1}^{C} n_{\cdot k}(\bar{X}_{\cdot\cdot k}-\bar{X}_{\cdots})^2$	MS_C/MS_e	$\sigma_e^2 + \sigma_\beta^2$
Row x Col	(R-1)(C-1)	$\sum_{j=1}^{R}\sum_{k=1}^{C}(\bar{X}_{\cdot jk}-\bar{X}_{\cdot j\cdot}-\bar{X}_{\cdot\cdot k}+\bar{X}_{\cdots})^2$	MS_{RC}/MS_e	$\sigma_e^2 + \sigma_{\alpha\beta}^2$
Error	$\sum_{j=1}^{R}\sum_{k=1}^{C}(n_{jk}-1)$	$\sum_{j=1}^{R}\sum_{k=1}^{C}\sum_{i=1}^{n_{jk}}(X_{ijk}-\bar{X}_{\cdot jk})^2$		σ^2
Total	N-1	$\sum_{j=1}^{R}\sum_{k=1}^{C}\sum_{i=1}^{n_{jk}}(X_{ijk}-\bar{X}_{\cdots})^2$		

where X_{ijk} is the score for individual i in Row j and column k,
$\bar{X}_{\cdot j\cdot}$ is the mean of row j,
$\bar{X}_{\cdot\cdot k}$ is the mean of column k,
$\bar{X}_{\cdot jk}$ is the mean of the cell for row j and column k,
$\bar{X}_{\cdots}$ is the overall (grand) mean.

As before, computational formulas may be developed from the defining formulas obtained from partitioning the total sum of squared deviations about the grand mean:

$$SS_T = \sum_{j=1}^{R}\sum_{k=1}^{C}\sum_{i=1}^{n_{jk}} X_{ijk} - T_{\cdots}^2/N \tag{5.4}$$

$$SS_R = \sum_{j=1}^{R} T_{\cdot j\cdot}^2/n_{j\cdot} - T_{\cdots}^2/N \tag{5.5}$$

$$SS_C = \sum_{k=1}^{C} T^2_{..k}/n_{.k} - T^2_{...}/N \qquad (5.6)$$

$$SS_{RC} = \sum_{j=1}^{R}\sum_{k=1}^{C} T^2_{.jk}/n_{jk} - SS_R - SS_C - T^2_{...}/N \qquad (5.7)$$

$$\sum_{j=1}^{R}\sum_{k=1}^{C}\sum_{i=1}^{n_{jk}} X^2_{ijk} - \sum_{j=1}^{R}\sum_{k=1}^{C} T^2_{.jk}/n_{jk} \qquad (5.8)$$

where T ... is the total of all scores,
T.jk is the total of the scores in a cell defined by the j row and k column,
T.j. is the total of the scores in the jth row,
T..k is the total of the scores in the kth column,
N is the total number of scores,
nj. is the number of scores in the jth row,
n.k is the number of scores in the kth column,
njk is the number of scores in the cell of the jth row and kth column.

In completing a two-way ANOVA, the researcher should attempt to have the same number of subjects in each group. If the ratio of any two columns is the same across rows then the cell sizes are proportional and the analysis is still legitimate. If cell sizes are neither equal nor proportional, then the total sum of squares does not equal the sum of squares for rows, columns, interaction and error and the F tests do not represent independent tests of significance.

Stating the Hypotheses

The individual score of a subject (Xijk) may be considered to be the linear composite of the effect of the row level (αj), the effect of the column (βk), the interaction effect of row and column combined ($\alpha\beta jk$), the overall mean and random error, that is

$$X_{ijk} = \mu + \alpha_j + \beta_k + \alpha\beta_{jk} + e_{ijk} \qquad (5.9)$$

The null hypotheses for the main effects therefore may be stated either as

$$H_o : \mu_{1.} = \mu_{2.} = \ldots = \mu_{j.} = \ldots \mu_{R.} \text{ for all rows,} \qquad (5.10)$$

$$\text{or } H_o : \alpha_1 \ldots = \alpha_j = \ldots \alpha_R \text{ for all rows, and} \qquad (5.11)$$

Analysis of Variance: The Two-Way, Fixed-Effects Design

$$H_o : \mu_{.1} = \ldots = \mu_{.k} = \ldots = \mu_{.C} \text{ for all columns,} \quad (5.12)$$

$$\text{or } H_o : \beta_1 = \ldots = \beta_k = \ldots = \beta_C \text{ for all columns, and} \quad (5.13)$$

$$H_o : (\mu_{jk} - \mu_{j.} - \mu_{.k} + \mu_{..}) \text{ for all row and column combinations,} \quad (5.14)$$

$$\text{or } H_o : \alpha\beta_{11} = \ldots = \alpha\beta_{jk} = \ldots = \alpha\beta_{RC} \text{ for interactions.} \quad (5.15)$$

$$\text{Again, we note that } \sum_{j=1}^{R} \alpha_{j.} = 0, \ \sum_{k=1}^{C} \beta_{.k} = 0 \text{ and } \sum_{j=1}^{R}\sum_{k=1}^{C} \alpha\beta_{jk} = 0. \quad (5.16)$$

Interpreting Interactions

One may examine the means of cells in a two-way ANOVA using a plot such as illustrated in the figure below for our example of the teacher's research:

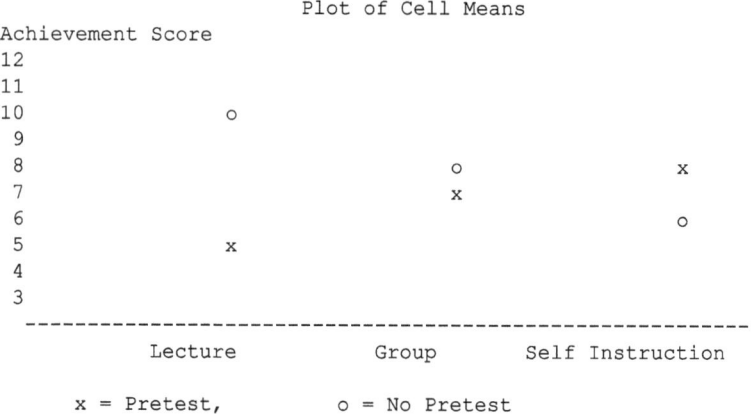

If lines are used to connect the o group means and lines are used to connect the x group means, one can see that the lines "cross".

If the lines for the pretest and no pretest levels are parallel across levels of the other factor, no interaction exists. When lines actually cross in the plot, this is called ordinal interaction. If the lines would cross if projected beyond current treatment levels, this is called disordinal interaction. In either case, the implication of interaction is that a particular combination of both treatments effects the dependent variable beyond the main effects alone. For example, if the interaction above is judged significant, then we cannot say that method 1 is better than method 3 of teaching without also specifying whether or not a pretest were used!

Note in the above interaction plot that the average of the three teaching method means are about the same for both pretest and no pretest conditions. This would indicate no main effect for the column variable pretest-no pretest. Similarly, the two means for each teaching method average about the same for each teaching method. This would indicate little effect of the variable teaching method (row). Your plot can graphically present effects due to the main variables as well as there interaction!

Random Effects Models

The two-way ANOVA design discussed to this point has assumed both factors contain fixed levels of treatment such that if the experiment was repeated numerous times, the levels would always be the same. If one or both of the factors represent random variables, that is, variables which would contain random levels upon replications of the experiment, then the expected values of the MSrows, MScolumns, and MSinteraction differ from that of the fixed-effects model. Presented below is a summary of the expected values for the two-way design when both variables are fixed, one variable random, and both variables random.

Both Row and Column Variables Fixed

Source	Expected MS	Calculated F-ratio
Row	$\sigma^2_e + nj.\sigma^2_\alpha$	MSR/MSe
Column	$\sigma^2_e + n.k\,\sigma^2_\beta$	MSC/MSe
Interaction	$\sigma^2_e + n_{jk}\sigma^2_{\alpha\beta}$	MSRC/MSe
Error	σ^2_e	

Rows Fixed, Columns Random

Source	Expected MS	Calculated F-ratio
Row	$\sigma^2_e + n..\sigma^2_{\alpha\beta} + nj\,\sigma^2_\alpha$	MSR/MSRC
Column	$\sigma^2_e + n.k\,\sigma^2_\beta$	MSC/MSe
Interaction	$\sigma^2_e + n..\sigma^2_{\alpha\beta}$	MSRC/MSe
Error	σ^2_e	

Row Random, Column Random

Source	Expected MS	Calculated F-ratio
Row	$\sigma^2_e + n..\sigma^2_{\alpha\beta} + n_j\sigma^2_\alpha$	MSR/MSRC
Column	$\sigma^2_e + n..\sigma^2_{\alpha\beta} + n_k.\sigma^2_\beta$	MSC/MSRC
Interaction	$\sigma^2_e + n..\sigma^2_{\alpha\beta}$	MSRC/MSe
Error	σ^2_e	

One Between, One Repeated Design

Introduction

A common research design in education involves repeated measurements of several groups of subjects. For example, a pre- and post test administered to students in experimental and control courses may be considered a mixed design with one between subjects factor and one within subjects (repeated measures) factor. We might hypothesize that the means of the pretest equals the posttest, hypothesize that the experimental and control group means are equal and hypothesize that the change from pretest to post-test is the same for the two groups. Tests for these hypotheses use the F statistic.

As another example, suppose we are interested in the teacher evaluations given by three groups of administrators before and after three different teacher-evaluation training programs. All administrators are provided identical information on a sample of teachers including level and content of courses taught, school characteristics, community and student characteristics, and teacher characteristics such as degree, years experience, professional memberships, etc. plus a videotape of teaching excerpts. Each subject reviews all information and makes teacher ratings. The subjects are then randomly assigned to the three treatments: (a) a program on teacher evaluation which stresses the motivational aspects, (b) a program which stresses the teacher improvement aspect and (c) a program which stresses the reward aspect. Following these programs, each subject again evaluates the same or parallel teachers. The hypotheses tested would be that the mean teacher evaluations of each experimental group are equal, the mean evaluations prior to programs equal mean evaluations following the programs, and the change in mean teacher evaluations from pre to post program time are equal.

The Research Design

The figure below presents the schema for the mixed between and within factors design. Note that the different subjects in each "A" treatment group are repeatedly measured under each of the "B" treatment conditions. Our sample size is n subjects in each of M groups and the number of treatments is L.

The main hypotheses to be tested are

$$H_0 : \mu_{1.} = \mu_{2.} = \ldots = \mu_{M.} \quad \text{(all A levels are equal).}$$
$$H_0 : \mu_{.1} = \mu_{.2} = \ldots = \mu_{.L} \quad \text{(all B levels are equal).}$$
$$H_0 : \mu_{11} = \mu_{jk} = \ldots = \mu_{ML} \quad \text{(all AB cells are equal).}$$

	B FACTOR TREATMENT LEVEL					Mean
	1	2	3	4	L	
A	X_{111}	X_{112}	X_{113}	X_{114}	X_{11L}	$\bar{X}_{11.}$
F	X_{211}	X_{212}	X_{213}	X_{214}........	X_{21L}	$\bar{X}_{21.}$
A C T O R	.	.	.	.	.	$\bar{X}_{.1.}$
	X_{i11}	X_{i12}	X_{i13}	X_{i14}........	X_{i1L}	$\bar{X}_{i1.}$
G R O U P	.	.	.	.	.	.
	X_{n11}	X_{n12}	X_{n13}	X_{n14}........	X_{n1L}	$\bar{X}_{n1.}$
1 Mean	$\bar{X}_{.11}$	$\bar{X}_{.12}$	$\bar{X}_{.13}$	$\bar{X}_{.14}$........	$\bar{X}_{.1L}$	$\bar{X}_{.1.}$
	X_{121}	X_{122}	X_{123}	X_{124}........	X_{12L}	$\bar{X}_{12.}$
	X_{221}	X_{222}	X_{223}	X_{224}........	X_{22L}	$\bar{X}_{22.}$
G R O U P	.	.	.	.	.	$\bar{X}_{.2.}$
	X_{i21}	X_{i22}	X_{i23}	X_{i24}........	X_{i2L}	$\bar{X}_{i2.}$
2	.	.	.	.	.	.
	X_{n21}	X_{n22}	X_{n23}	X_{n24}........	X_{n2L}	$\bar{X}_{n2.}$
Mean	$\bar{X}_{.21}$	$\bar{X}_{.22}$	$\bar{X}_{.23}$	$\bar{X}_{.24}$........	$\bar{X}_{.2L}$	$\bar{X}_{.2.}$
Col. Means	$\bar{X}_{..1}$	$\bar{X}_{..2}$	$\bar{X}_{..3}$	$\bar{X}_{..4}$........	$\bar{X}_{..L}$	$\bar{X}_{...}$

Theoretical Model

The theoretical model for a subject i's score X from group j in Factor A on treatment k from factor B may be written

$$X_{ijk} = \mu + \alpha_j + \beta_k + \pi_{i(j)} + \beta_{jk} + \beta\pi_{ki(j)} + e_{i(jk)} \tag{5.17}$$

where μ is the population mean of the scores,
αj is the effect of treatment j in Factor A,
ßk is the effect of treatment k in Factor B,
πi is the effect of person i,
αßjk is the interaction of Factor A treatment j and treatment level k in Factor B,
ßπ ki(j) is the interaction of subject i and B treatment k in the jth treatment group of A,
and ei(jk) is the error for person i in treatment j of Factor A and treatment k of Factor B.

In an experiment, we are usually interested in estimating the effect size of each treatment in each factor. We may also be interested in knowing whether or not there are significant differences among the subjects, and whether or not different subjects react differently to various treatments.

Assumptions

As in most ANOVA designs, we make a number of assumptions. For the mixed factors design these are:

1. The sum of treatment effects (α_j) is equal to zero,
2. The sum of treatment effects (ßk) is equal to zero,
3. The sum of person effects ($\pi i(j)$) is equal to zero,
4. The sum of αßjk interaction effects is equal to zero,
5. The sum of ßαki(j) interaction effects is equal to zero,
6. The sum of treatment x person interaction effects within levels of A (ßπki(j)) is zero,
7. The errors (ei(jk)) are normally distributed with mean zero,
8. The variance of errors in each A treatment (α_j) are equal,
9. The variance of errors in each B treatment (βk) are equal,
10. The covariances among the treatments (COVpq(j) p<>q p,q = 1..L) within j levels of A are all equal.

The last assumption, equal covariances, means that if we were to transform scores within treatments to z scores, the correlations among the scores between any two treatments would all be equal in the population. You will also note that the denominator of the F ratios for testing differences among A treatment means is the pooled variance among subject means within groups as in a one-way ANOVA and the denominator of the F statistic for the Factors of B (the repeated measures) and the A × B interaction F statistic is the variance due to the pooled treatment by subjects interaction found in the Treatments by Subjects design.

Summary Table

The AxS ANOVA Summary table is often presented as follows:

Source	D.F.	SS	MS	F
Between subjects	$Mn - 1$	$\sum_{j=1}^{M}\sum_{i=1}^{n} L(X_{ij.} - X_{...})^2$		
A	$M - 1$	$nL\sum_{j=1}^{M} (\bar{X}_{.j.} - \bar{X}_{...})^2$	$SSA/(M - 1)$	MSA/MSSwG
Subjects within groups	$M(n - 1)$	$\sum_{j=1}^{M}\sum_{i=1}^{n} L(\bar{X}_{ij.} - \bar{X}_{.j.})^2$	$SSSwG/[M(n - 1)]$	
Within subjects	$Mn(L - 1)$	$\sum_{j=1}^{M}\sum_{k=1}^{L}\sum_{i=1}^{n}(\bar{X}_{ijk} - \bar{X}_{.jk})^2$		
B	$L - 1$	$nM\sum_{k=1}^{L} (\bar{X}_{..k} - \bar{X}_{...})^2$	$SSB/(L - 1)$	MSB/MSBxSwG
A × B	$(M - 1)(L - 1)$	$n\sum_{j=1}^{M}\sum_{k=1}^{L}(\bar{X}_{.jk} - \bar{X}_{..k} - \bar{X}_{.j.} + \bar{X}_{...})^2$	$SSAxS/[(M - 1)(L - 1)]$	
B × S within groups	$M(n - 1)(L - 1)$	$\sum_{j=1}^{M} SS_{BS(j)}$	$SSBxSwG/[M(n - 1)(L - 1)]$	
Total	$nML - 1$	$\sum_{j=1}^{M}\sum_{k=1}^{L}\sum_{i=1}^{n}(\bar{X}_{ijk} - \bar{X}_{..})^2$		

Population Parameters Estimated

The population mean of all scores (μ) is estimated by the overall mean. The mean squares provide estimates as follows:

$$MS_A \text{ estimates } \sigma_e^2 + M\sigma_\pi^2 + Mn\sigma_\alpha^2$$

$$MS_{SwG} \text{ estimates } \sigma_e^2 + M\sigma_\pi^2$$

$$MS_B \text{ estimates } \sigma_e^2 + \sigma_{\beta\pi}^2 + Mn\sigma_\beta^2$$

$$MS_{AB} \text{ estimates } \sigma_e^2 + \sigma_{\beta\pi}^2 + n\sigma_{\alpha\beta}^2$$

$$MS_{BxSwG} \text{ estimates } \sigma_e^2 + \sigma_{\beta\pi}^2$$

Two Factor Repeated Measures Analysis

Repeated measures designs have the advantage that the error terms are typically smaller that designs using independent groups of observations. This was true for the Student t-test using matched or correlated scores. On the down-side, repeated measures on the same objects pose a special problem, particularly when the objects are human subjects. The main problem is "practice" or "learning" effects that may be greater for one treatment level than another. These effects are completely confounded with the actual treatment effects. While random or counter-balanced assignment of the treatments may reduce the cumulative effects to some degree, it does not remove the effects specific to a given treatment. It is also assumed that the covariance matrices are equal among the treatment levels. Users of these designs with human subjects should be careful to minimize the practice effects. This can sometimes be done by having subjects do tasks that are similar to those in the actual experiment before beginning trials of the experiment.

Nested Factors Analysis of Variance Design

The Research Design

In the Nested ANOVA design, one factor (B) is completely nested within levels of another factor (A). Thus unlike the A × B Fixed Effects ANOVA in which all levels of B are crossed with all levels of A, each level of B is found in only one level of A in the nested design. The design may be graphically depicted as below:

A Factor	Treatment 1		Treatment j	Treatment M	
B Factor	Level 1	Level 2	... Level k	.... Level L	
Obser-	X_{111}	X_{112}	.... X_{1jk}	.... X_{1ML}	
vations	X_{211}	X_{212}	.... X_{2jk}	.... X_{2ML}	
	.	.	.... .	.... .	
	.	.	.... .	.... .	
	X_{n11}	X_{n12}	.... X_{njk}	.... X_{nML}	
B Means	$\bar{X}_{.11}$	$\bar{X}_{.12}$	.... $\bar{X}_{.jk}$	.... $\bar{X}_{.ML}$	
A Means	$\bar{X}_{.1.}$		$\bar{X}_{.j.}$	$\bar{X}_{.M.}$	

The Variance Model

The observed X scores may be considered to be composed of several effects:

$$X_{ijk} = \mu + \alpha_j + \beta_{k(j)} + e_{k(j)} \tag{5.18}$$

The ANOVA Summary Table

We partition the total squared deviations of X scores from the grand mean of scores into sources of variation. The independent sources may be used to form F ratios for the hypothesis that the treatment means of A levels are equal and the hypothesis that the treatment levels of B are equal. The summary table (with sums of squares derivations) is as follows:

Source	D.F.	SS	Estimates:
A[a]	$M - 1$	$\sum_{j=1}^{M} n_{j.}(\bar{X}_{.j.} - \bar{X}_{...})^2$	$\sigma^2_e + nD\sigma^2_B + nM\sigma^2_\alpha$
B (pooled)	$\sum_{j=1}^{M}(q_j - 1)$	$\sum_{j=1}^{M}\sum_{k=1}^{L_j} n_{jk}(\bar{X}_{.jk} - \bar{X}_{.j.})^2$	$\sigma^2_e + n\sigma^2_B$
Within	$\sum_{j=1}^{M}\sum_{k=1}^{L}(n_{jk} - 1)$	$\sum_{j=1}^{M}\sum_{k=1}^{L_j}\sum_{i=1}^{n_{jk}}(\bar{X}_{ijk} - \bar{X}_{.jk})^2$	σ^2_e
Total	$N - 1$	$\sum_{j=1}^{M}\sum_{k=1}^{L_j}\sum_{i=1}^{n_{jk}}(\bar{X}_{ijk} - \bar{X}_{...})^2$	

[a]Note: When factor B is a random effect, $D = 1$ and the F ratio for testing the A effect is the MSA/MSB. When factor B is a fixed effect, $D = 0$ and the F ratio for testing A effects is MSA/MSw.
where:
Xijk = An observed score in B treatment level k under A treatment level j,
X.jk = the mean of observations in B treatment level k in A treatment level j,
X.j. = the mean of observations in A treatment level j,
X... = the grand mean of all observations,
njk = the number of observations in B treatment level k under A treatment level j
nj. = the number of observations in A treatment level j,
N = the total number of observations.

A, B and C Factors with B Nested in A

$$\text{MODEL}: \bar{X}_{ijk} = \mu + \alpha_i + \beta_{j(i)} + \gamma_k + \alpha\gamma_{ik} + \beta\gamma_{jk} + \varepsilon_{ijk} \tag{5.19}$$

Assume that an experiment involves the use of two different teaching methods, one which involves instruction for 1 consecutive hour and another that involves two half-hours of instruction 4 h apart during a given day. Three schools are randomly selected to provide method 1 and three schools are selected to provide method 2. Note that school is *nested* within method of instruction. Now assume that n subjects are randomly selected for each of two categories of students in each school. Category 1 students are males and category 2 students are female. This design may be illustrated in the table below:

	Instruction method 1			Instruction method 2		
	School 1	School 2	School 3	School 4	School 5	School 6
Category 1	n	n	n	n	n	n
Category 2	n	n	n	n	n	n

Notice that without School, the Categories are crossed with method and therefore are NOT nested. The expected values of the mean squares is:

Source of variation	df	Expected value
A (Method)	$p - 1$	$\sigma^2_e + nD_qD_r\sigma^2_{\beta\gamma} + nqD_r\sigma^2_{\alpha\gamma} + nrD_q\sigma^2_\beta + nqr\sigma^2_\alpha$
B within A	$p(q - 1)$	$\sigma^2_e + nD_r\sigma^2_{\beta\gamma} + nr\sigma^2_\beta$
C (Category)	$r - 1$	$\sigma^2_e + nD_q\sigma^2_{\beta\gamma} + nqD_p\sigma^2_{\alpha\gamma} + npq\sigma^2_\gamma$
AC	$(p - 1)(r - 1)$	$\sigma^2_e + nD_q\sigma^2_{\beta\gamma} + nq\sigma^2_{\alpha\gamma}$
(B within A)C	$p(q - 1)(r - 1)$	$\sigma^2_e + n\sigma^2_{\beta\gamma}$
Within cell	$pqr(n - 1)$	σ^2_e

where there are p methods of A, q nested treatments B (Schools) and r C treatments (Categories). The D's with subscripts q, r or p have the value of 0 if the source is fixed and a value of 1 if the source is random. In this version of the analysis, all effects are considered fixed (D's are all zero) and therefore the F tests all use the Within Cell mean square as the denominator. If you use random treatment levels, you may need to calculate a more appropriate F test.

Latin and Greco-Latin Square Designs

> Did you hear about the ancient roman statistician who was always called a nerd? Turns out he was just a Latin Square.

Some Theory

In a typical two or three-way analysis of variance design, there are independent groups assigned to each combination of the A, B (and C) treatment levels. For example, if one is designing an experiment with three levels of Factor A, four levels of Factor B and two levels of Factor C, then a total of 24 groups of randomly selected subjects would be used in the experiment (with random assignment of the groups to the treatment combinations.) With only four observations (subjects) per group, this would require 96 subjects in total. In such a design, one can obtain the main effects of A, B and C independent of the A × B, A × C, B × C and A × B × C interaction effects of the treatments. Often however, one may know before hand by previous research or by logical reasoning that the interactions should be minimal or would not exist. When such a situation exists, one can use a design which confounds or partially confounds such interactions with the main effects and drastically reduces the number of treatment groups required for the analysis. If the subjects can be repeatedly observed under various treatment conditions as in some of the previously discussed repeated-measures designs, then one can even further reduce the number of subjects required in the experiment. The designs to be discussed in this section utilize what are known as "Latin Squares".

The Latin Square

A Latin square is a balanced two-way classification scheme. In the following arrangement of letters, each letter occurs just once in each row and once in each column:

A	B	C
B	C	A
C	A	B

If we interchange the first and second row we obtain a similar arrangement with the same characteristics:

B	C	A
A	B	C
C	A	B

Two Latin squares are orthogonal if, when they are combined, the same pair of symbols occurs no more than once in the composite squares. For example, if the two Latin squares labeled Factor A and Factor B are combined to produce the composite shown below those squares the combination is NOT orthogonal because treatment combinations A1B2, A2B3, and A3B1 occur in more than one cell. However, if we combine Factor A and Factor C we obtain a combination that IS orthogonal.

FACTOR A			FACTOR B			FACTOR C		
A1	A2	A3	B2	B3	B1	C1	C2	C3
A2	A3	A1	B3	B1	B2	C3	C1	C2
A3	A1	A2	B1	B2	B3	C2	C3	C1

COMBINED A and B
A1B2 A2B3 A3B1
A2B3 A3B1 A1B2
A3B1 A1B2 A2B3

COMBINED A and C
A1C1 A2C2 A3C3
A2C3 A3C1 A1C2
A3C2 A1C3 A2C1

Notice that the three levels of treatment A and the three levels of treatment C are combined in such a way that no one combination is found in more than one cell. When two Latin squares are combined to form an orthogonal combination of the two treatment factors, the combination is referred to as a Greco-Latin square. Notice that the number of levels of both the treatment factors must be the same to form a square. Extensive tables of orthogonal Latin squares have been compiled by Cochran and Cox in "Experimental Designs", New York, Wiley, 1957.

Typically, the Greco-Latin square is represented using only the number (subscripts) combinations such as:

11 22 33
23 31 12
32 13 21

One can obtain additional squares by interchanging any two rows or columns of a Greco-Latin square. Not all Latin squares can be combined to form a Greco-Latin square. For example, there are no orthogonal squares for 6 by 6 or for 10 by 10 Latin squares. If the dimensions of a Latin square can be expressed as a prime number raised to the power of any integer n, then orthogonal squares exist. For example, orthogonal Latin squares exist of dimension 3, 4, 5, 8 and 9 from the relationships 3 from 3^1, 4 from 2^2, 5 from 5^1, 8 from 2^3, 9 from 3^2, etc.

Latin squares are often tabled in only "standard form". A square in standard form is one in which the letters of the first row and column are in sequence. For example, the following is a standard form for a four dimension square:

A B C D
B A D C
C D B A
D C A B

There are potentially a large number of standard forms for a Latin square of dimension n. There are 4 standard forms for a 4 by 4 square, and 9,408 standard forms for a 6 by 6 square. By interchanging rows and columns of the standard forms, one can create additional non-standard forms. For a 4 by 4 there are a total of

576 Latin squares and for a 6 by 6 there are a total of 812,851,200 squares! One can select at random a standard form for his or her design and then randomly select rows and columns to interchange to create a randomized combination of treatments.

Plan 1 by B. J. Winer

In his book "Statistical Principles in Experimental Design", New York, McGraw-Hill, 1962, Winer outlines a number of experimental designs that utilize Latin squares. He refers to these designs as "Plans" 1 through 13 (with some variations in several plans.) Not all plans have been included in OpenStat. Eight have been selected for inclusion at this time. The most simple design is that which provides the following model and estimates:

$$\text{MODEL}: X_{ijkm} = \mu + \alpha_{i(s)} + \beta_{j(s)} + \gamma_{k(s)} + \text{res}_{(s)} + \varepsilon_{m(ijk)} \tag{5.20}$$

Where i, j, k refer to levels of Factors A, B and C and m the individual subject in the unit. The (s) indicates this is a model from a Latin (s)quare design.

Source of variation	Degrees of freedom	Expected mean square
A	p − 1	$\sigma^2_\varepsilon + np\sigma^2_\alpha$
B	p − 1	$\sigma^2_\varepsilon + np\sigma^2_\beta$
C	p − 1	$\sigma^2_\varepsilon + np\sigma^2_\gamma$
Residual	(p − 1)(p − 2)	$\sigma^2_\varepsilon + np\sigma^2_{res}$
Within cell	p2(n − 1)	σ^2_ε

In the above, p is the dimension of the square and n is the number of observations per unit.

Plan 2

Winer's Plan 2 expands the design of Plan 1 discussed above by adding levels of a Factor D. Separate Latin Squares are used at each level of Factor D. The plan of the design might appear as below:

		Factor B					Factor B		
		B1	B2	B3			B1	B2	B3
	Factor					Factor			
Factor D1	A1	C3	C2	C1	Factor D2	A1	C1	C3	C2
	A2	C1	C3	C2		A2	C2	C1	C3
	A3	C2	C1	C3		A3	C3	C2	C1

Latin and Greco-Latin Square Designs

The analysis of Plan 2 is as follows:

Source of variation	Degrees of freedom	Expected mean square
A	p − 1	$\sigma^2_\varepsilon + npq\sigma^2_\alpha$
B	p − 1	$\sigma^2_\varepsilon + npq\sigma^2_\beta$
C	p − 1	$\sigma^2_\varepsilon + npq\sigma^2_\gamma$
D	q − 1	$\sigma^2_\varepsilon + npq\sigma^2_\delta$
AD	(p − 1)(q − 1)	$\sigma^2_\varepsilon + npq\sigma^2_{\alpha\delta}$
BD	(p − 1)(q − 1)	$\sigma^2_\varepsilon + npq\sigma^2_{\beta\delta}$
CD	(p − 1)(q − 1)	$\sigma^2_\varepsilon + npq\sigma^2_{\gamma\delta}$
Residual	q(p − 1)(p − 2)	$\sigma^2_\varepsilon + npq\sigma^2_{res}$
Within cell	$p^2 q(n − 1)$	σ^2_ε

Notice that we can obtain the interactions with the D factor since all A, B and C treatments in the Latin square are observed under each level of D. The model for Plan 2 expected value of the observed (X) score is:

$$X_{ijkmo} = \mu + \alpha_{i(s)} + \beta_{j(s)} + \gamma_{k(s)} + \delta_m + \alpha\delta_{i(s)m} + \beta\delta_{j(s)m} + \gamma\delta_{k(s)m} + \text{res}_{(s)} \quad (5.21)$$

As in Plan 1 described above, the (s) indicates sources from the Latin square.

Plan 3 Latin Squares Design

Plan 3 utilizes a balanced set of p × p Latin squares in a p × p × p factorial experiment. An example for a 3 × 3 × 3 design is shown below:

	Factor B				Factor B				Factor B		
	B1	B2	B3		B1	B2	B3		B1	B2	B3
Factor				Factor				Factor			
Factor D1 A1	C1	C2	C3	Factor D2 A1	C2	C3	C1	Factor D3 A1	C3	C1	C2
A2	C2	C3	C1		A2	C3	C1	C2		A2	C1	C2	C3
A3	C3	C1	C2		A3	C1	C2	C3		A3	C2	C3	C1

The levels of factors A, B and C are assigned at random to the symbols defining the Latin square. The levels of factor D are assigned at random to the whole squares. Notice the levels of each factor must be p, unlike the previous plan 2. In a complete four factor design with three levels of each factor there would be 81 cells however with this design there are only 27. The main effect of factor D will be partially confounded with the ABC interaction however the main effects of A, B and C as well as the their interactions will be complete. The model of this design is:

$$E(X_{ijkmo}) = \mu + \alpha_i + \beta_j + \gamma_k + \alpha\beta_{ij} + \alpha\gamma_{ik} + \beta\gamma_{jk} + \delta_m + \alpha\beta\gamma'_{ijk} \quad (5.22)$$

The sources of variation, their degrees of freedom and parameter estimates are as shown below:

Source	D.F.	E(MS)
A	$p - 1$	$\sigma^2_\varepsilon + np^2\sigma^2_\alpha$
B	$p - 1$	$\sigma^2_\varepsilon + np^2\sigma^2_\beta$
C	$p - 1$	$\sigma^2_\varepsilon + np^2\sigma^2_\gamma$
AB	$(p - 1)(p - 1)$	$\sigma^2_\varepsilon + np\sigma^2_{\alpha\beta}$
AC	$(p - 1)(p - 1)$	$\sigma^2_\varepsilon + np\sigma^2_{\alpha\gamma}$
BC	$(p - 1)(p - 1)$	$\sigma^2_\varepsilon + np\sigma^2_{\beta\gamma}$
D	$p - 1$	$\sigma^2_\varepsilon + np^2\sigma^2_\delta$
(ABC)'	$(p - 1)^3 - (p - 1)$	$\sigma^2_\varepsilon + n\sigma^2_{\alpha\beta\gamma}$
Within cell	$p^3(n - 1)$	σ^2_ε

Analysis of Greco-Latin Squares

A Greco-Latin square design permits a three-way control of experimental units (row, column, and layer effects) through use of two Latin squares that are combined. One square is denoted with Latin letters and the other with Greek letters as illustrated below:

Square I			Square II			Combined squares		
A	B	C	α	β	γ	Aα	Bβ	Cγ
B	C	A	γ	α	β	Bγ	Cα	Aβ
C	A	B	β	γ	α	Cβ	Aγ	Bα

Using numbers for the levels of the first and second effects, the composite square might also be represented by:

$$\begin{array}{ccc} 11 & 22 & 33 \\ 23 & 31 & 12 \\ 32 & 13 & 21 \end{array}$$

There are actually four variables: row, column, Latin-letter and Greek letter variables with p-squared cells in the composite square rather than p * p * p * p as there would be in a four-factor factorial design. The main effects of each of the factors will be confounded with the two-factor and higher interaction effects. Therefore this design is limited to the situations where the four factors are assumed to have negligible interactions. It is assumed that there are n independent observations in each cell.

Latin and Greco-Latin Square Designs

The analysis that results provides the following sources of variation:

Source	D.F.	E(MS)
A (Rows)	p − 1	$\sigma^2_\varepsilon + np\sigma^2_\alpha$
B (Columns)	p − 1	$\sigma^2_\varepsilon + np\sigma^2_\beta$
C (Latin letters)	p − 1	$\sigma^2_\varepsilon + np\sigma^2_\gamma$
D (Greek letters)	p − 1	$\sigma^2_\varepsilon + np\sigma^2_\delta$
Residual	(p − 1)(p − 3)	$\sigma^2_\varepsilon + n\sigma^2_{res}$
Within cell	p2(n − 1)	σ^2_ε
Total	np2 − 1	

Plan 5 Latin Square Design

When the same unit (e.g. subject) may be observed under different treatment conditions, a considerable saving is realized in the sample size necessary for the experiment. As in all repeated measures designs however one must make certain assumptions about the homogeneity of variance and covariance. In plan 5 the levels of treatment under factor B are arranged in a Latin square with the columns representing levels of factor A. The rows are groups of subjects for which repeated measures are made across the columns of the square. The design is represented below:

		Factor A levels		
		A1	A2	A3
Group	G1	B3	B1	B2
	G2	B1	B2	B3
	G3	B2	B3	B1

The model of the analysis is:

$$E(X_{ijkm}) = \mu + \delta_k + \pi_{m(k)} + \alpha_i + \beta_j + \alpha\beta'_{ij} \qquad (5.23)$$

The sources of variation are estimated by:

Source	D.F.	E(MS)
Between subjects	np − 1	
B	p − 1	$\sigma^2_\varepsilon + p\sigma^2_\pi + np\sigma^2_\delta$
Subjects in groups	p(n − 1)	$\sigma^2_\varepsilon + p\sigma^2_\pi$
Within subjects	np(p − 1)	
A	p − 1	$\sigma^2_\varepsilon + np\sigma^2_\alpha$
B	p − 1	$\sigma^2_\varepsilon + np\sigma^2_\beta$
(AB')	(p − 1)(p − 2)	$\sigma^2_\varepsilon + n\sigma^2_{\alpha\beta}$
Error (within)	p(n − 1)(p − 1)	σ^2_ε

Plan 6 Latin Squares Design

Winer indicates that Plan 6 may be considered "as a fractional replication of a three-factor factorial experiment arranged in incomplete blocks." Each subject within Group 1 is assigned to to treatement combinations abc$_{111}$, abc$_{231}$ and abc$_{321}$ such that each subject in the group is observed under all levels of factors A and B but under only one level of factor C. There is no balance with respect to any of the interactions but there is balance with respect to factors A and B. If all interactions are negligible relative to the main effects the following model and the sources of variation are appropriate:

$$E(X_{ijkm}) = \mu + \gamma_{k(s)} + \pi_{m(k)} + \alpha_{i(s)} + \beta_{j(s)} + res_{(s)}. \qquad (5.24)$$

Source of variation	D.F.	E(MS)
Between subjects	np − 1	
C	p − 1	$\sigma^2_\varepsilon + p\sigma^2_\pi + np\sigma^2_\gamma$
Subjects within groups	p(n − 1)	$\sigma^2_\varepsilon + p\sigma^2_\pi$
Within subjects	np(p − 1)	
A	p − 1	$\sigma^2_\varepsilon + np\sigma^2_\alpha$
B	p − 1	$\sigma^2_\varepsilon + np\sigma^2_\beta$
Residual	(p − 1)(p − 2)	$\sigma^2_\varepsilon + n\sigma^2_{res}$
Error (within)	p(n − 1)(p − 1)	σ^2_ε

The experiment may be viewed (for three levels of each variable) in the design below:

| Group | Levels of C | Levels of factor A | | |
		A1	A2	A3
G1	C1	B1	B3	B2
G2	C2	B2	B1	B3
G3	C3	B3	B2	B1

Plan 7 for Latin Squares

If, in the previous plan 6 we superimpose the Factors B and C as orthogonal Latin Squares, then Factor C is converted into a within-subjects effect. The Greco-Latin square design may be viewed as the following (for three levels of treatment):

| Group | Levels of factor A | | |
	A1	A2	A3
G1	BC11	BC23	BC32
G2	BC22	BC31	BC13
G3	BC33	BC12	BC21

The expected value of X is given as:

$$E(X_{ijkmo}) = \mu + \delta_{m(s)} + \pi_{o(m)} + \alpha_{i(s)} + \beta_{j(s)} + \gamma_{k(s)} \qquad (5.25)$$

The sources of variation, their degrees of freedom and the expected mean squares are:

Source of variation	D.F.	E(MS)
Between subjects	np − 1	
Groups	p − 1	$\sigma^2_\varepsilon + p\sigma^2_\pi + np\sigma^2_\delta$
Subjects within groups	p(n − 1)	$\sigma^2_\varepsilon + p\sigma^2_\pi$
Within subjects	np(p − 1)	
A	p − 1	$\sigma^2_\varepsilon + np\sigma^2_\alpha$
B	p − 1	$\sigma^2_\varepsilon + np\sigma^2_\beta$
C	p − 1	$\sigma^2_\varepsilon + np\sigma^2_\gamma$
Residual	(p − 1)(p − 3)	$\sigma^2_\varepsilon + n\sigma^2_{res}$
Error (within)	p(n − 1)(p − 1)	σ^2_ε

Plan 9 Latin Squares

If we utilize the same Latin square for all levels of a Factor C we would have a design which looks like the outline shown below for three levels:

Levels of factor C											
C1				C2				C3			
Levels of factor A				Levels of factor A				Levels of factor A			
Group	A1	A2	A3	Group	A1	A2	A3	Group	A1	A2	A3
G1	B2	B3	B1	G4	B2	B3	B1	G7	B2	B3	B1
G2	B1	B2	B3	G5	B1	B2	B3	G8	B1	B2	B3
G3	B3	B1	B2	G6	B3	B1	B2	G9	B3	B1	B2

The model for expected values of X is:

$$E(X_{ijkmo}) = \mu + \gamma_k + (\text{row})_m + (\gamma \times \text{row})_{km} + \pi_{o(m)} + \alpha_i + \beta_j + \alpha\beta'_{ij}$$
$$+ \alpha\gamma_{ik} + \beta\gamma_{jk} + \alpha\beta\gamma'_{ijk} \qquad (5.26)$$

The sources of variation for Plan 9 are shown below:

Source of variation	D.F.	E(MS)
Between subjects	npq − 1	
C	q − 1	$\sigma^2_\varepsilon + p\sigma_2 + np^2\sigma^2_\gamma$

(continued)

Source of variation	D.F.	E(MS)
Rows [AB(between)]	$p - 1$	$\sigma^2_\varepsilon + p\sigma_2 + nq\sigma^2_{\alpha\beta}$
C x row [ABC(between)]	$(p - 1)(q - 1)$	$\sigma^2_\varepsilon + p\sigma_2 + n\sigma^2_{\alpha\beta}$
Subjects within groups	$pq(n - 1)$	$\sigma^2_\varepsilon + p\sigma_2$
Within subjects	$npq(p - 1)$	
A	$p - 1$	$\sigma^2_\varepsilon + npq\sigma^2_\alpha$
B	$p - 1$	$\sigma^2_\varepsilon + npq\sigma^2_\beta$
AC	$(p - 1)(q - 1)$	$\sigma^2_\varepsilon + np\sigma^2_{\alpha\gamma}$
BC	$(p - 1)(q - 1)$	$\sigma^2_\varepsilon + np\sigma^2_{\beta\gamma}$
(AB)'	$(p - 1)(p - 2)$	$\sigma^2_\varepsilon + nq\sigma^2_{\alpha\beta}$
(ABC)'	$(p - 1)(p - 3)(q - 1)$	$\sigma^2_\varepsilon + n\sigma^2_{\alpha\beta\gamma}$
Error (within)	$pq(p - 1)(n - 1)$	σ^2_ε

In this design the groups and subjects within groups are considered random while, like previous designs, the A,B and C factors are fixed. Interactions with the group and subject effects are considered negligible.

Analysis of Variance Using Multiple Regression Methods

A Comparison of ANOVA and Regression

In one-way analysis of variance with Fixed Effects, the model that describes the expected Y score is usually given as

$$Y_{i,j} = \mu + \alpha_j + e_{i,j} \tag{5.27}$$

where $Y_{i,j}$ is the observed dependent variable score for subject i in treatment group j,
μ is the population mean of the Y scores,
α_j is the effect of treatment j, and
$e_{i,j}$ is the deviation of subject i in the jth treatment group from the population mean for that group.

The above equation may be rewritten with sample estimates as

$$Y'_{i,j} = \bar{Y}_{..} + (\bar{Y}_{.j} - \bar{Y}_{..}) \tag{5.28}$$

For any given subject then, irrespective of group, we have

$$Y'_i = \bar{Y}_{..} + (\bar{Y}_{.1} - \bar{Y}_{..})X_1 + \ldots + (\bar{Y}_{.k} - \bar{Y}_{..})X_k \tag{5.29}$$

Analysis of Variance Using Multiple Regression Methods

where X_j is 1 if the subject is in the group, otherwise 0.

If we let $B_0 = Y..$ and the effects $(Y_{.j} - Y_{..})$ be B_j for any group, we may rewrite the above equation as

$$Y'_i = B_0 + B_1 X_1 + \ldots + B_k X_k \quad (5.30)$$

This is, of course, the general model for multiple regression! In other words, the model used in ANOVA may be directly translated to the multiple regression model. They are essentially the same model!

You will notice that in this model, each subject has K predictors X. Each predictor is coded a 1 if the subject is in the group, otherwise 0. If we create a variable for each group however, we do not have independence of the predictors. We lack independence because one group code is redundant information with the $K - 1$ other group codes. For example, if there is only two groups and a subject is in group 1, then $X_1 = 1$ and X_2 MUST BE 0 since an individual cannot belong in both groups. There are only $K - 1$ degrees of freedom for group membership – if an individual is not in groups 1 up to K we automatically know they belong to the Kth group. In order to use multiple regression, the predictor variables must be independent. For this reason, the number of predictors is restricted to one less than the number of groups. Since all α_j effects must sum to zero, we need only know the first $K - 1$ effects – the last can be obtained by subtraction from $1 - \Sigma \alpha_j$ where $j = 1,\ldots,K - 1$.

We also remember that

$$B_0 = \bar{Y}.. - (B_1 \bar{X}_1 + \ldots + B_k \bar{X}_k). \quad (5.31)$$

Effect Coding

In order for B_0 to equal the grand mean of the Y scores, we must restrict our model in such a way that the sum of the products of the X means and regression coefficients equals zero. This may be done by use of "effect" coding. In this method there are $K - 1$ independent variables for each subject. If a subject is in the group corresponding to the jth variable, he or she has a score $X_j = 1$ otherwise the score is $X_j = 0$. Subjects in the Kth group do not have a corresponding X variable so they receive a score of 1 in all of the group codes.

As an example, assume that you have five subjects in each of three groups. The "effect" coding of predictor variables would be

SUBJECT	Y	CODE 1	CODE 2	
01	5	1	0	
02	8	1	0	
03	4	1	0	(Group 1)
04	7	1	0	
05	3	1	0	
06	4	0	1	
07	6	0	1	
08	2	0	1	(Group 2)
09	9	0	1	
10	4	0	1	
11	3	−1	−1	
12	6	−1	−1	
13	5	−1	−1	(Group 3)
14	9	−1	−1	
15	4	−1	−1	

You may notice that the mean of X_1 and of X_2 are both zero. The cross-products of $X_1 X_2$ is n_3, the size of the last group.

If we now perform a multiple regression analysis as well as a regular ANOVA for the data above, we will obtain the following results:

SOURCE	DF	SS	MS	F	PROB>F
Full Model	2	0.533	0.267	0.048	0.953
Groups	2	0.533	0.267	0.048	0.953
Residual	12	66.400	5.533		
Total	14	66.933			

$R^2 = 0.008$

You will note that the SS_{groups} may be obtained from either the ANOVA printout or the SS_{reg} in the Multiple Regression analysis. The SS_{error} is the same in both analyses as is the total sum of squares.

Orthogonal Coding

While effect coding provides the means of directly estimating the effect of membership in levels or treatment groups, the correlations among the independent variables are not zero, thus the inverse of that matrix may be difficult if done by hand. Of greater interest however, is the ability of other methods of data coding that permits the research to pre-specify contrasts or comparisons among particular treatment groups of interest. The method of orthogonal coding has several benefits:

Analysis of Variance Using Multiple Regression Methods

1. The user can pre-plan comparisons among selected groups or treatments, and
2. The inter-correlation matrix is a diagonal matrix, that is, all off-diagonal values are zero. This results in a solution for the regression coefficients which can easily be calculated by hand.

When orthogonal coding is utilized, there are $K - 1$ possible orthogonal comparisons in each factor. For example, if there are four treatment levels of Factor A, there are 3 possible orthogonal comparisons that may be made among the treatment means. To illustrate orthogonal coding, we will utilize the same example as before. The previous effect coding will be replaced by orthogonal coding as illustrated in the data below:

SUBJECT	Y	CODE 1	CODE 2	
01	5	1	1	
02	8	1	1	
03	4	1	1	(Group 1)
04	7	1	1	
05	3	1	1	
06	4	-1	1	
07	6	-1	1	
08	2	-1	1	(Group 2)
09	9	-1	1	
10	4	-1	1	
11	3	0	-2	
12	6	0	-2	
13	5	0	-2	(Group 3)
14	9	0	-2	
15	4	0	-2	

Now notice that, as before, the sum of the values in each coding vector is zero. Also note that, in this case, the product of the coding vectors is also zero. (Multiply the code values of two vectors for each subject and add up the products – they should sum to zero.) Vector 1 above (Code 1) represents a comparison of treatment group 1 with treatment group 2. Vector 2 represents a comparison of groups 1 AND 2 with group 3.

Now let us look at coding for, say, five treatment groups. The coding vectors below might be used to obtain orthogonal contrasts:

GROUP	VECTOR 1	VECTOR 2	VECTOR 3	VECTOR 4
1	1	1	1	1
2	-1	1	1	1
3	0	-2	1	1
4	0	0	-3	1
5	0	0	0	-4

As before, the sum of coefficients in each vector is zero and the product of any two vectors is also zero. This assumes that there are the same number of subjects in each group. If groups are different in size, one may use additional multipliers based on the proportion of the total sample found in each group. The treatment group number in the left column may, of course, represent any one of the treatment groups thus it is possible to select a specific comparison of interest by assigning the treatment groups in the order necessary to obtain the comparison of interest.

Return now to the previous example. The results from the regression analysis program as well as the ANOVA program are presented in the figures below. The first figure presents the inter-correlation matrix among the variables. Notice that the inter-correlations among the coding vectors are zero. The next figure presents the R^2 and the summary of regression coefficients. Multiplication of the R^2 times the sum of squares for the dependent variable will yield the sum of squares for regression. This will equal the sum of squares for groups in the subsequent ANOVA results table. By use of orthogonal vectors, we may also note that the regression coefficients are simply the correlation of each vector with the dependent variable. Multiplication of the squared regression coefficients times the sum of squares total will therefore give the sum of squares due to each contrast. The total sum of squares for groups is simply the sum of the sum of squares for each contrast! The test of departure of the regression coefficients from zero is a test of significance for the contrast in the corresponding coding vector. The a priori specified contrasts, unlike post-hoc comparisons maintain the selected alpha rate and more power. Hence, sensitivity to true population treatment effects are more likely to be detected by the planned comparison than by a post-hoc comparison.

Dummy Coding

Effect and orthogonal coding methods both resulted in code vectors which summed to zero across the subjects. In each of those cases, the constant B_0 estimates the population mean since it is the grand mean of the sample (see equation 5.3). Both methods of coding also resulted in the same squared multiple correlation coefficient R^2 indicating that the proportion of variance explained by both methods is the same.

Another method of coding which is popular is called "dummy" coding. In this method, $K - 1$ vectors are also created for the coding of membership in the K treatment groups. However, the sum of the coded vectors do not add to zero as in the previous two methods. In this coding scheme, if a subject is a member of treatment group 1, the subject receives a code of 1. All other treatment group subjects receive a code of 0. For a second vector (where there are more than two treatment groups), subjects that are in the second treatment group are coded with a 1 and all other treatment group subjects are coded 0. This method continues for the $K - 1$ groups. Clearly, members of the last treatment group will have a code of zero in all vectors. The coding of members in each of five treatment groups is illustrated below:

```
GROUP    VECTOR 1      VECTOR 2      VECTOR 3      VECTOR 4

  1          1             0             0             0
  2          0             1             0             0

  3          0             0             1             0

  4          0             0             0             1

  5          0             0             0             0
```

With this method of coding, like that of effect coding, there will be correlations among the coding vectors which differ from zero thus necessitating the computation of the inverse of a symmetric matrix rather than a diagonal matrix. Never the less, the squared multiple correlation coefficient R^2 will be the same as with the other coding methods and therefore the SS_{reg} will again reflect the treatment effects. Unfortunately, the resulting regression coefficients reflect neither the direct effect of each treatment or a comparison among treatment groups. In addition, the constant B_0 reflects the mean only of the treatment group (last group) which receives all zeroes in the coding vectors. If however, the overall effects of treatment is the finding of interest, dummy coding will give the same results.

Two Factor ANOVA by Multiple Regression

In the above examples of effect, orthogonal and dummy coding of treatments, we dealt only with levels of a single treatment factor. We may, however, also analyze multiple factor designs by multiple regression using each of these same coding methods. For example, a two-way analysis of variance using two treatment factors will typically provide the test of effects for the A factor, the B factor and the interaction of the A and B treatments. We will demonstrate the use of effect, orthogonal and dummy coding for a typical research design involving three levels of an A treatment and four levels of a B treatment.

```
                       Example Design

                   Levels of Treatment B

                  |   1   |   2   |   3   |   4   |
                  |_____|_____|_____|_____|
   Levels    1    |       |       |       |       |
                  |_____|_____|_____|_____|
    of       2    |       |       |       |       |
                  |_____|_____|_____|_____|
 Treatment   3    |       |       |       |       |
                  |_____|_____|_____|_____|
     A
```

For effect coding in the above design, we apply effect codes to the A treatment levels first and then, beginning again, to the B treatment levels independently of the A codes. Finally, we multiple each of the code vectors of the A treatments times each of the code vectors of the B treatment to create the interaction vectors. The vectors below illustrate this for the above design:

		A		B			A x B					
		X_1	X_2	X_3	X_4	X_5	X_6	X_7	X_8	X_9	X_{10}	X_{11}
ROW	COL	A_1	A_2	B_1	B_2	B_3	A_1B_1	A_1B_2	A_1B_3	A_2B_1	A_2B_2	A_2B_3
1	1	1	0	1	0	0	1	0	0	0	0	0
1	2	1	0	0	1	0	0	1	0	0	0	0
1	3	1	0	0	0	1	0	0	1	0	0	0
1	4	1	0	-1	-1	-1	-1	-1	-1	0	0	0
2	1	0	1	1	0	0	0	0	0	1	0	0
2	2	0	1	0	1	0	0	0	0	0	1	0
2	3	0	1	0	0	1	0	0	0	0	0	1
2	4	0	1	-1	-1	-1	0	0	0	-1	-1	-1
3	1	-1	-1	1	0	0	-1	0	0	-1	0	0
3	2	-1	-1	0	1	0	0	-1	0	0	-1	0
3	3	-1	-1	0	0	1	0	0	-1	0	0	-1
3	4	-1	-1	-1	-1	-1	1	1	1	1	1	1

If you add the values in any one of the vectors above you will see they sum to zero. In addition, the product of any two vectors selected from a combination of treatment A, B or A × B sets will also be zero! With effect coding, the treatment effect vectors from one factor are orthogonal (uncorrelated) with the treatment effect vectors of the other factor as well as the interaction effect vectors. The effect vectors within each treatment or interaction are not, however, orthogonal.

With effect coding, we may "decompose" the R^2 for the full model into the three separate parts, that is

$$R^2_{y.1\,2\,3\,4\,5\,6\,7\,8\,9\,10\,11} = R^2_{y.1\,2} + R^2_{y.3\,4\,5} + R^2_{y.6\,7\,8\,9\,10\,11} \tag{5.32}$$

since the A, B and A × B effects are orthogonal.

Again, the regression coefficients directly report the effect of treatment group membership, that is, B_1 is the effect of treatment group 1 in the A factor and B_2 is the effect of treatment group 2 in the A factor. The effect of treatment group 3 in the A factor can be obtained as

$$\alpha_3 = 1 - \Sigma(\alpha_1 + \alpha_2) = 1 - (B_1 + B_2) \quad (5.33)$$

since the sum of effects is constrained to equal zero. Similarly, B_3 estimates β_1, B_4 estimates β_2 and B_5 estimates the B factor effect β_3 for column 3. The effect of column four is also obtained as before, that is,

$$\beta_4 = 1 - (B_3 + B_4 + B_5). \quad (5.34)$$

The interaction effects for the cells, $\alpha\beta_{ij}$, may be obtained from the regression coefficients corresponding to the interaction vectors. In this example, B_6 estimates $\alpha\beta_{11}$, B_7 estimates $\alpha\beta_{12}$, B_8 estimates $\alpha\beta_{13}$, B_9 estimates $\alpha\beta_{21}$, B_{10} estimates $\alpha\beta_{22}$ and B_{11} estimates $\alpha\beta_{23}$. Since the sum of the interaction effects in any row or column must be zero, we can determine estimates for the cells in rows 1 and 2 of column 4 as follows:

$$\alpha\beta_{14} = 1 - (B_6 + B_7 + B_8) \quad \text{and} \quad (5.35)$$

$$\alpha\beta_{24} = 1 - (B_9 + B_{10} + B_{11}). \quad (5.36)$$

We may also utilize orthogonal coding vectors within each treatment factor as we did for effect coding above. The same two-factor design above could utilize the vectors below:

		A		B				A x B			
		X1	X2	X3	X4	X5	X6	X7	X8	X9	X10	X11
ROW	COL	A1	A2	B1	B2	B3	A1B1	A1B2	A1B3	A2B1	A2B2	A2B3
1	1	1	1	1	1	1	1	1	1	1	1	1
1	2	1	1	-1	1	1	-1	1	1	-1	1	1
1	3	1	1	0	-2	1	0	-2	1	0	-2	1
1	4	1	1	0	0	-3	0	0	-3	0	0	-3
2	1	-1	1	1	1	1	-1	-1	-1	1	1	1
2	2	-1	1	-1	1	1	1	-1	-1	-1	1	1
2	3	-1	1	0	-2	1	0	2	-1	0	-2	1
2	4	-1	1	0	0	-3	0	0	3	0	0	-3

3	1	0	-2	1	1	1	0	0	0	-2	-2	-2
3	2	0	-2	-1	1	1	0	0	0	2	-2	-2
3	3	0	-2	0	-2	1	0	0	0	0	4	-2
3	4	0	-2	0	0	-3	0	0	0	0	0	6

As before, the sum of each vector is zero. This time however, the product of vectors within each factor as well as between factors and interaction are zero. All vectors are orthogonal to one another. The inter-correlation matrix is therefore a diagonal matrix and easily inverted by hand. The R^2 for the full model may be easily decomposed into the sum of squared simple correlations between the dependent and independent score vectors, that is

$$\begin{aligned} R^2_{y.1\,2\,3\,4\,5\,6\,7\,8\,9\,10\,11} \\ = r^2_{y.1} + r^2_{y.2} + & \quad \text{(row effects)} \\ r^2_{y.3} + r^2_{y.4} + r^2_{y.5} + & \quad \text{(column effects)} \\ r^2_{y.6} + r^2_{y.7} + r^2_{y.8} + r^2_{y.9} + r^2_{y.10} + r^2_{y.11} & \quad \text{(interaction effects)} \end{aligned} \quad (5.37)$$

The regression coefficients obtained with orthogonal coding vectors represent planned comparisons among treatment means. Using the coding vectors for this example, the B_1 coefficient would represent the comparison of row 1 mean with row 2 mean. B_2 would represent the contrast of row 3 mean with the combination of rows 1 and 2. The coefficients B_3, B_4 and B_5 similarly contrast column means. The contrasts represented by the interaction vectors will reflect comparisons among specific cell combinations. For example, B_7 above will reflect a contrast of the combined cells in row 1 column 1 and row 2 column 2 with the combined cells of row 1 column 2 and row 2 column 1.

Analysis of Covariance by Multiple Regression Analysis

In the previous sections we have examined methods for coding nominal variables of analysis of variance designs to explain the variance of the continuous dependent variable. We may, however, also include one or more independent variables that are continuous and expected to have the same correlation with the dependent variable in each treatment group population. As an example, assume that the two-way ANOVA design discussed in the previous section represents an experiment in which Factor A represent three type of learning reinforcement (positive only, negative only and combined positive and negative) while Factor B represents four types of learning situations (CAI, teacher led, self instruction, and peer tutor). Assume the dependent variable is a standardized measure of Achievement in learning the French language. Finally, assume the treatment groups are exposed to the treatments for a sufficiently long period of time to produce measurable achievement by most students and that the students have been randomly assigned to the treatment groups. It may occur to the reader that achievement in learning a new language might be related to general intelligence as measured, say, by the Stanford-Binet Intelligence Test as well as related to prior English achievement measured by a standardized achievement test in English. Variation in IQ and English achievement of subjects in the treatment groups may explain a portion of the within treatment cell variance. We prefer to have the within cell variance as small as possible since it is the basis of the mean squared residual used in the F tests of our treatment effects. To accomplish this, we can first extract that portion of total dependent score variance explained by IQ and English achievement before examining that portion of the remaining variance explainable by our main treatment effects. Assume therefore, that in addition to the 11 vectors representing Factor A level effects, Factor B level effects and Factor interaction effects, we include X_{12} and X_{13} predictors of IQ and English. Then the proportion of variance for Factor A effects controlling for IQ and English is

$$R^2_{y \cdot 1\ 2\ 3\ 4\ 5\ 6\ 7\ 8\ 9\ 10\ 11\ 12\ 13} - R^2_{y \cdot 3\ 4\ 5\ 6\ 7\ 8\ 9\ 10\ 11\ 12\ 13}$$

The proportion of French achievement variance due to Factor B treatments controlling for IQ and English would be

$$R^2_{y \cdot 1\ 2\ 3\ 4\ 5\ 6\ 7\ 8\ 9\ 10\ 11\ 12\ 13} - R^2_{y \cdot 1\ 2\ 6\ 7\ 8\ 9\ 10\ 11\ 12\ 13}$$

and the proportion of variance due to interaction of Factor A and Factor B controlling for IQ and English would be

$$R^2_{y \cdot 1\ 2\ 3\ 4\ 5\ 6\ 7\ 8\ 9\ 10\ 11\ 12\ 13} - R^2_{y \cdot 1\ 2\ 3\ 4\ 5\ 12\ 13}$$

In each of the above, the full model contains all predictors while the restricted model contains all variables except those of the effects being evaluated. The F statistic for testing the hypothesis of equal treatment effects is

$$F = \frac{R^2_{full} - R^2_{restricted}}{1.0 - R^2_{full}} \cdot \frac{N - K_f - 1}{K_f - K_r} \qquad (5.38)$$

where K_f is the number of predictors in the full model, and K_r is the number of predictors in the restricted model.

The numerator and denominator degrees of freedom for these F statistics is $(K_f - K_r)$ and $(N - K_f - 1)$ respectively.

Analysis of Covariance assumes homogeneity of covariance among the treatment groups (cells) in the populations from which the samples are drawn. If this assumption holds, the interaction of the covariates with the main treatment factors (A and B in our example) should not account for significant variance of the dependent variable. You can explicitly test this assumption therefore by constructing a full model which has all of the previously included independent variables plus prediction vectors obtained by multiplying each of the treatment level vectors times each of the covariates. In our above example, for instance, we would multiply each of the first five vectors times both IQ and English vectors (X_{12} and X_{13}) resulting in a full model with ten more variables (23 predictors in all).

The R^2 from our previous full model would be subtracted from the R^2 for this new full model to determine the proportion of variance attributable to heteroscedasticity of the covariance among the treatment groups. If the F statistic for this proportion is significant, we cannot employ the analysis of covariance model. The implication would be that somehow, IQ and prior English achievement interacts differently among the levels of the treatments. Note that in testing this assumption of homogeneity of covariance, we have a fairly large number of variables in the regression analysis. To obtain much power in our F test, we need a considerable number of subjects. Several hundred subjects would not be unreasonable for this study, i.e. 25 subjects per each of the eight treatment groups!

Sums of Squares by Regression

The General Linear Model

We have seen in the above discussion that the multiple regression method may be used to complete an analysis of variance for a single dependent variable. The model for multiple regression is:

$$y_i = \sum_{j=1}^{k} B_j X_j + e_i \qquad (5.39)$$

where the jth B value is a coefficient multiplied times the jth independent predictor score, Y is the observed dependent score and e is the error (difference between the observed and the value predicted for Y using the sum of weighted independent scores).

In some research it is desirable to determine the relationship between multiple dependent variables and multiple independent variables. Of course, one could complete a multiple regression analysis for each dependent variable but this would ignore the possible relationships among the dependent variables themselves. For example, a teacher might be interested in the relationship between the sub-scores on a standardized achievement test (independent variables) and the final examination results for several different courses (dependent variables.) Each of the final examination scores could be predicted by the sub-scores in separate analyses but most likely the interest is in knowing how well the sub-scores account for the combined variance of the achievement scores. By assigning weights to each of the dependent variables as well as the independent variables in such a way that the composite dependent score is maximally related to the composite independent score we can quantify the relationship between the two composite scores. We note that the squared product-moment correlation coefficient reflects the proportion of variance of a dependent variable predicted by the independent variable.

We can express the model for the general linear model as:

$$YM = BX + E \qquad (5.40)$$

where Y is an n (the number of subjects) by m (the number of dependent variables) matrix of dependent variable values, M is a m by s (number of coefficient sets), X is a n by k (the number of independent variables) matrix, B is a k by s matrix of coefficients and E is a vector of errors for the n subjects.

The General Linear Model (GLM) procedure is an analysis procedure that encompasses a variety of analyses. It may incorporate multiple linear regression as well as canonical correlation analysis as methods for analyzing the user's data. In some commercial statistics packages the GLM method also incorporates non-linear analyses, maximum-likelihood procedures and a variety of tests not found in the current version of this model. The version in OpenStat is currently limited to a single dependent variable (continuous measure.) You should complete analyses with multiple dependent variables with the Canonical Correlation procedure.

One can complete a variety of analyses of variance with the GLM procedure including multiple factor ANOVA and repeated and mixed model ANOVAs.

The output of the GLM can be somewhat voluminous in that the effects of treatment variables and covariates are analyzed individually by comparing regression models with and without those variables.

Canonical Correlation

Introduction

Canonical correlation analysis involves obtaining an index that describes the degree of relationship between two variables, each of which is a weighted composite of other variables. We have already examined the situation of an index between one variable and a weighted composite variable when we studied the multiple correlation coefficient of chapter 4. Using a form similar to that used in multiple regression analysis, we might consider:

$$\beta_{y1}Y_1 + \beta_{y2}Y_2 + .. + \beta_{ym}Y_m + \beta_y = \beta_{x1}X_1 + .. + \beta_{xn}X_n + \beta_x$$

as a model for the regression of the composite function Y_c on the composite function Xc where

$$Y_c = \sum_{i=1}^{m} \beta_{yi} Y_i \quad \text{and} \quad X_c = \sum_{j=1}^{n} \beta_{yj} X_j \qquad (5.41)$$

and the Y and X scores are in standardized form (z scores).

Using "least-squares" criteria, we can maximize the simple product-moment correlation between Yc and Xc by selecting coefficients (Betas) which minimize the residuals (e). As in multiple regression, this involves solving partial derivatives for the β's on each side of the equation. The least-squares solution is more complicated than for multiple regression and will not be covered in this text. (See T. W. Anderson, An Introduction to Multivariate Analysis, 1958, chapter 12).

Unfortunately for the beginning student, the canonical correlation analysis does not yield just one correlation index (R_c), but in fact may yield up to m or n (whichever is smaller) independent coefficients. This is because there are additional linear functions of the X's and Y's which may "explain" the residual variances $._y$ and $._x$ not explained by the first set of βx and βy weights. Each set of these canonical functions explains an additional portion of the common variance of the X and Y variables!

Before introducing the mathematics of obtaining these canonical correlations, the sets of corresponding weights and statistical tests of significance, we need to have a basic understanding of the concept of roots and vectors of a matrix.

Eigenvalues and Eigenvectors

A concept which occurs frequently in multivariate statistical analyses is the concept of eigenvalues (roots) and associated eigenvectors. Canonical correlation, factor analysis, multivariate analysis of variance, discriminant analysis, etc. utilize the roots and vectors of matrices in their solutions. To understand this concept, consider a k by k matrix (e.g. a correlation matrix)$[R]_{kxk}$. A basic problem in mathematical statistics is to find a $k \times 1$ vector (matrix) $[E]_j$ and a scalar (single value) y_j such that

$$[R]_{kxk}[E]_{kx1} = y_j [E]_{kx1} \quad \text{where at least one element} \qquad (5.42)$$
$$\text{of } [E]_{kx1} \text{ is not zero.}$$

This equation may be rewritten as

$$[R]_{kxk} [E]_{kx1} - y_j [E]_{kx1} = [0]_{kx1}$$

$$\text{or as} \quad \left([R]_{kxk} - y_j [I]_{kxk}\right) [E]_{kx1} = [0]_{kx1} \qquad (5.43)$$

If we were to solve this equation for [E] by multiplying both sides of the last equation by the inverse of the matrix in the parenthesis (assuming the inverse exists), then [E] would be zero, a solution which violates our desire that at least one element of [E] NOT be zero! Consequently, [E] will have a non-zero element only if the determinant of

$$\left([R]_{kxk} - y_j [I]_{kxk} \right)$$

is zero. The equation

$$\left| [R]_{kxk} - y_j [I]_{kxk} \right| = 0 \tag{5.44}$$

is called the characteristic equation. The properties of this equation have many applications in science and engineering.

The vector $[E]_{kx1}$ and the scalar y_j in the (5.43) are the eigenvector and eigenvalue of the matrix $[R]_{kxk}$.

Eigenvalues and eigenvectors are also known as characteristic roots and vectors of a matrix. To demonstrate that the eigenvalue is a root of a characteristic equation, consider the simple case of a 2 × 2 matrix such as

$$\begin{vmatrix} b_{11} & b_{12} \\ b_{21} & b_{22} \end{vmatrix}$$

The problem is to find the root y_j in solving

$$\begin{vmatrix} b_{11} & b_{12} \\ b_{21} & b_{22} \end{vmatrix} \cdot \begin{vmatrix} e_1 \\ e_2 \end{vmatrix} = y_j \begin{vmatrix} e_1 \\ e_2 \end{vmatrix}$$

Using the determinant:

$$\begin{vmatrix} b_{11} & b_{12} \\ b_{21} & b_{22} \end{vmatrix} - \begin{vmatrix} y & 0 \\ 0 & y \end{vmatrix} = 0$$

or

$$\begin{vmatrix} b_{11}-y & b_{12} \\ b_{21} & b_{22}-y \end{vmatrix} = 0$$

This determinant has the solution

$$(b_{11} - y)(b_{22} - y) - b_{12}b_{21} = 0$$

or $b_{11}b_{22} - yb_{22} - yb_{11} + y^2 - b_{12}b_{21} = 0$

or $y^2 - y(b_{22} + b_{11}) + (b_{11}b_{22} - b_{12}b_{21}) = 0$

This is a quadratic equation with two roots y_1 and y_2 given by

$$.5 \left\{ (b_{22} + b_{11}) + / - \left[(b_{22} + b_{11})^2 - 4(b_{11}b_{22} - b_{12}b_{21}).5 \right] \right\}$$

With the roots y_1 and y_2 evaluated, the elements e_1 and e_2 of the eigenvector can be solved from

$$\begin{vmatrix} b_{11} & b_{12} \\ b_{21} & b_{22} \end{vmatrix} \begin{vmatrix} e_1 \\ e_2 \end{vmatrix} = y_j \begin{vmatrix} e_1 \\ e_2 \end{vmatrix}$$

which reduces to the equations (for each root):

$$b_{11}e_1 + b_{12}e_2 = y\, e_1$$
$$b_{21}e_1 + b_{22}e_2 = y\, e_2$$

and further reduces to

$$(b_{11} - y)e_1 + b_{12}e_2 = 0$$
$$b_{21}e_1 + (b_{22} - y)e_2 = 0$$

Solving these last equations simultaneously for e_1 and e_2 will yield the elements of the eigenvector [E].

There will be an eigenvector for each eigenvalue. In the case of the 2 × 2 matrix, the complete solution will be

$$\begin{vmatrix} b_{11} & b_{12} \\ b_{21} & b_{22} \end{vmatrix} \begin{vmatrix} e_{11} & e_{12} \\ e_{21} & e_{22} \end{vmatrix} = \begin{vmatrix} y_1 & 0 \\ 0 & y_2 \end{vmatrix} \begin{vmatrix} e_{11} & e_{12} \\ e_{21} & e_{22} \end{vmatrix} \quad (5.45)$$

Every k × k matrix will have as many eigenvalues and eigenvectors as its order. Not all of the eigenvalues may be nonzero. When a square matrix [R] is symmetric, its eigenvalues are all real and the associated eigenvectors are orthogonal (products equal zero). If some of the eigenvalues are zero, we say that the RANK of the matrix is (k − p) where p is the number of roots equal to zero. The TRACE of a symmetric matrix is the sum of the eigenvalues. The determinant of the matrix is the product of all roots.

Other relationships obtainable from symmetric matrices are:

$$[R]_{kxk} [E]_{kxk} = [y]_{kxk} [E]_{kxk} \qquad (5.46)$$

$$c[R]_{kxk} [E]_{kxk} = c[y]_{kxk} [E]_{kxk} \quad \text{where c is a constant.}$$

It may be pointed out that for any symmetric matrix and its eigenvalues there may be an infinite number of associated eigenvector matrices. There is, however, at least one matrix of eigenvectors that is *orthonormal*. An orthonormal matrix is one which when premultiplied by its transpose yields an identity matrix. If [E] is orthonormal then:

$$[E]'_{kxk} [E]_{kxk} = [I]_{kxk} \qquad (5.47)$$

$$\text{and} \quad [E]'_{kxk} = [E]^{-1}_{kxk} \qquad (5.48)$$

Did you hear about the statistician who was looking all over for the sum of eigenvalues from a variance–covariance matrix but couldn't find a trace?

The Canonical Analysis

In completing a canonical analysis, the inter-correlation matrix among all of the variables may be partitioned into four sub-matrices as shown symbolically below. The $[R_{11}]$ matrix is the matrix of correlations among the "left_hand" variables of the equation presented earlier. The $[R_{22}]$ matrix is the correlations among the "right_hand" variables of our model. $[R_{12}]$ are the inter-correlations among the left and right hand variables. $[R_{21}]$ is the transpose of $[R_{12}]$.

$$[R] = \begin{vmatrix} R_{11} & R_{12} \\ R_{21} & R_{22} \end{vmatrix} \qquad (5.49)$$

To start the canonical analysis, a product matrix is first formed by:

$$[R_p] = [R_{22}]^{-1} [R_{21}] [R_{11}]^{-1} [R_{12}] \qquad (5.50)$$

The equation

$$\Big([R_p] - y_j[I]\Big) v_j = 0 \qquad (5.51)$$

where y_j is the jth root and v_j is the corresponding eigenvector is solved using the characteristic equation:

$$\left|[R_p] - y_j[I]\right| = 0 \tag{5.52}$$

with the restriction that

$$[V]'[R_{22}][V] = [I] \tag{5.53}$$

The canonical correlation $_1R_c$ corresponding to the first linear relationship between the left hand variables and the right hand variables is equal to the square root of the first root y_1. In general, the jth canonical correlation is obtained as:

$$_jR_c = \sqrt{y_j} \tag{5.54}$$

The canonical correlation may be interpreted as the product-moment correlation between a weighted composite of the left-hand variables and a weighted composite of the right-hand variables.

Discriminant Function/MANOVA

Theory

Multiple discriminant function analysis is utilized to obtain a set of linear functions which maximally discriminate (differentiate) among subjects belonging to several different groups or classifications. For example, an investigator may want to develop equations which differentiate among successful occupational groups based on responses to items of a questionnaire. The functions obtained may be written as:

$$F_j = B_{j,1}X_1 + \ldots + B_{j,m}X_m \tag{5.55}$$

where

Xi represents an observed variable (i = 1..m),
Bj,i is a coefficient for the Xi variable from the
jth discriminant function

The coefficients of these discriminant functions are the normalized vectors corresponding to the roots obtained for the matrix

$$[P] = [W]^{-1}[A] \tag{5.56}$$

where

[W] − 1 is the inverse of the pooled within groups deviation score cross-products and

[A] is the among groups cross-products of deviations of group means from the grand mean (weighted by the group size).

Once the discriminant functions are obtained, they may be used to classify subjects on the basis of their continuous variables. The number of functions to be applied to each individual's set of X scores will be one less than the number of groups or the number of X variables (whichever is less). Subjects are then classified into the group for which their discriminant score has the highest probability of belonging.

Discriminant function analysis and Multivariate Analysis of Variance results are essentially identical. The Wilk's Lambda statistic, the Rao F statistic and the Bartlett Chi-Squared statistic will yield the same inference regarding significant differences among the groups. The discriminant functions may be used to obtain a plot of the subjects in the discriminant space, that is, the Cartesian (orthogonal) space of the discriminant functions. By examining these plots and the standardized coefficients which contribute the most to each discriminant function, you can determine those variables which appear to best differentiate among the groups.

Cluster Analyses

Theory

Objects or people may form groups on the basis of similarity of scores on one or more variables. For example, students in a school may form groups relatively homogeneous with regard to interests in music, athletics, science, languages, etc. An investigator may not have "a priori" groups but rather, be interested in identifying "natural" groupings based on similar score profiles. The Cluster programs of this chapter provide the capability of combining subjects which have the most similar profile of scores.

Hierarchical Cluster Analysis

This procedure was adapted from the Fortran program provided by Donald J. Veldman in his 1967 book. To begin, the sum of squared differences for each pair of subjects on K variables is calculated. If there are n subjects, there are n * (n − 1)/2 pairings. That pair of subjects yielding the smallest sum of squared differences is then combined using the average of the pair on each variable, forming a new "subject" or group. The process is repeated with a new combination formed each time. Eventually, of course, all subjects are combined into a single group. The decision as to when to stop further clustering is typically based on an "error" estimate which reflects the variability of scores for subjects in groups. As in analysis of variance, the between group variability should be significantly greater than the within group variability, if there are to be significant differences among the groups formed.

When you begin execution of the program, you are asked to identify the variables in your data file that are to be used in the grouping. You are also asked to enter the number of groups at which to begin printing the members within each cluster. This may be any value from the total number of subjects down to 2. In practice, you normally select the value of the "ideal" number of groups you expect or some slightly larger value so you can see the increase in error which occurs as more and more of the groups and subjects are combined into new groups. You may also specify the significance level necessary to end the grouping, for example, the value .05 is frequently used in one-way ANOVA analyses when testing for significance. The value used is in fact referred to the F distribution for an F approximation to a multivariate Wilk's Lambda statistic.

Path Analysis

Theory

Path analysis is a procedure for examining the inter-correlations among a set of variables to see if they are consistent with a model of causation. A causal model is one in which the observed scores (events) of an object are assumed to be directly or indirectly caused by one or more preceding events. For example, entrance to college may be hypothesized to be a result of high achievement in high school. High achievement in high school may be the result of parent expectations and the student's intelligence. Intelligence may be a result of parent intelligence, early nutrition, and early environmental stimulation, etc., etc. Causing and resultant variables may be described in a set of equations. Using standardized z scores, the above example might be described by the following equations:

1. $z_1 = e_1$ Parent intelligence
2. $z_2 = P_{21}z_1 + e_2$ Child's nutrition
3. $z_3 = P_{31}z_1 + P_{32}z_2 + e_3$ Child's intelligence
4. $z_4 = P_{41}z_1 + e_4$ Parent expectations
5. $z_5 = P_{53}z_3 + P_{54}z_4 + e_5$ School achievement
6. $z_6 = P_{63}z_3 + P_{64}z_4 + P_{65}z_5 + e_6$ College GPA

In the above equations, the P's represent path coefficients measuring the strength of causal effect on the resultant due to the causing variable z. In the above example, z1 has no causing variable and path coefficient. It is called an exogenous variable and is assumed to have only external causes unknown in this model. The "e" values represent contributions that are external and unknown for each variable. These external causes are assumed to be uncorrelated and dropped from further computations in this model. By substituting the definitions of each z score in a model like the above, the correlation between the observed variables can be expressed as in the following examples:

$$r_{12} = 3z_1z_2/n = P_{21}3z_1z_1/n = P_{21} \tag{5.57}$$

$$r_{23} = 3z_2z_3/n = P_{31}P_{21} + P_{32} \tag{5.58}$$

etc.

In other words, the correlations are estimated to be the sum of direct path coefficients and products of indirect path coefficients. The path coefficients are estimated by the standardized partial regression coefficients (betas) of each resultant variable on its causing variables. For example, coefficients P31 and P32 above would be estimated by ß31.2 and ß32.1 in the multiple regression equation

$$z_3 = \beta_{31.2}z_1 + \beta_{32.1}z_2 + e_3 \tag{5.59}$$

If the hypothesized causal flow model sufficiently describes the interrelationships among the observed variables, the reproduced correlation matrix using the path coefficients should deviate only by sampling error from the original correlations among the variables.

When you execute the Path Analysis procedure in OpenStat, you will be asked to specify the exogenous and endogenous variables in your analysis. The program then asks you to specify, for each resultant (endogenous) variable, the causing variables. In this manner you specify your total path model. The program then completes the number of multiple regression analyses required to estimate the path coefficients, estimate the correlations that would be obtained using the model path coefficients and compare the reproduced correlation matrix with the actual correlations among the variables.

You may discover in your reading that this is but one causal model method. More complex methods include models involving latent variables (such as those identified through factor analysis), correlated errors, adjustments for reliability of the variables, etc. Structural model equations of these types are often analyzed using the LISREL™ package found in commercial packages such as SPSS™ or SAS™.

Factor Analysis

The Linear Model

Factor analysis is based on the procedure for obtaining a new set of uncorrelated (orthogonal) variables, usually fewer in number than the original set, that reproduces the co-variability observed among a set or original variables. Two models are commonly utilized:

1. The principal components model wherein the observed score of an individual i on the jth variable Xi,j is given as:

$$X_{i,j} = A_{j,1}S_{i,1} + A_{j,2}S_{i,2} + \ldots + A_{j,k}S_{i,k} + C \qquad (5.60)$$

where Aj,k is a loading of the kth factor on variable j,
Si,k is the factor score of the ith individual on the kth factor and
C is a constant.

The Aj,k loadings are typically least-squares regression coefficients.

2. The common factor model assumes each variable X may contain some unique component of variability among subjects as well as components in common with other variables. The model is:

$$X_{i,j} = A_{j,1}S_{i,1} + \ldots + A_{j,k}S_{i,k} + A_{j,u}S_{i,u} \qquad (5.61)$$

The above equation may also be expressed in terms of standard z scores as:

$$z_{i,j} = a_{j,1}S_{i,1} + \ldots + a_{j,k}S_{i,k} + a_{j,u}S_{i,u} \qquad (5.62)$$

Since the average of standard score products for the n cases is the product-moment correlation coefficient, the correlation matrix among the j observed variables may be expressed in matrix form as:

$$[R]_{jxj} = [F]_{jxk} [F]'_{kxj} - [U]^2_{jxj} \qquad \text{(array sizes k< = j)} \qquad (5.63)$$

The matrix [F] is the matrix of factor loadings or correlations of the k theoretical orthogonal variables with the j observed variables. The [U] matrix is a diagonal matrix with unique loadings on the diagonal.

The factor loadings above are the result of calculating the eigenvalues and associated vectors of the characteristic equation:

$$\left| [R] - [U]^2 - [I] \right| \qquad (5.64)$$

where the lambda values are eigenvalues (roots) of the equation.

When you execute the Factor Analysis Program in OpenStat, you are asked to provide information necessary to complete an analysis of your data file. You enter the name of your file and identify the variables to analyze. If you elect to send output to the printer, be sure the printer is on when you start. You will also be asked to specify the type of analysis to perform. The principle components method, a partial image analysis, a Guttman Image Analysis, a Harris Scaled Image Analysis, a Canonical Factor Analysis or an Alpha Factor Analysis may be elected. Selection of the method depends on the assumptions you make concerning sampling of variables and sampling of subjects as well as the theory on which you view your variables. You may request a rotation of the resulting factors which follows completion of the analysis of the data,. The most common rotation performed is the Varimax rotation. This method rotates the orthogonal factor loadings so that the loadings within each factor are most variable on the average. This tends to produce "simple structure", that is, factors which have very high or very low loadings for the original variables and thus simplifies the interpretation of the resulting factors. One may also elect to perform a Procrustean rotation whereby the obtained factor loadings are rotated to be maximally congruent with another factor loading matrix. This second set of loadings which is entered by the user is typically a set which represents some theoretical structure of the original variables. One might, however, obtain factor loadings for males and females separately and then rotate one solution against the other to see if the structures are highly similar for both sexes.

Chapter 6
Non-Parametric Statistics

Beginning statistics students are usually introduced to what are called "parametric" statistics methods. Those methods utilize "models" of score distributions such as the normal (Gaussian) distribution, Poisson distribution, binomial distribution, etc. The emphasis in parametric statistical methods is estimating population parameters from sample statistics when the distribution of the population scores can be assumed to be one of these theoretical models. The observations made are also assumed to be based on continous variables that utilize an interval or ratio scale of measurement. Frequently the measurement scales available yield only nominal or ordinal values and nothing can be assumed about the distribution of such values in the population sampled. If however, random sampling has been utilized in selecting subjects, one can still make inferences about relationships and differences similar to those made with parametric statistics. For example, if students enrolled in two courses are assigned a rank on their achievement in each of the two courses, it is reasonable to expect that students that rank high in one course would tend to rank high in the other course. Since a rank only indicates order however and not "how much" was achieved, we cannot use the usual product–moment correlation to indicate the relationship between the ranks. We can estimate, however, what the product of rank values in a group of n subjects where the ranks are randomly assigned would tend to be and estimate the variability of these sums or rank products for repeated samples. This would lead to a test of significance of the departure of our rank product sum (or average) from a value expected when there is no relationship.

A variety of non-parametric methods have been developed for nominal and ordinal measures to indicate congruence or similarity among independent groups or repeated measures on subjects in a group.

Contingency Chi-Square

The frequency chi-square statistic is used to accept or reject hypotheses concerning the degree to which observed frequencies depart from theoretical frequencies in a row by column contingency table with fixed marginal frequencies. It therefore tests the independence of the categorical variables defining the rows and columns. As an example, assume 50 males and 50 females are randomly assigned to each of three types of instructional methods to learn beginning French, (a) using a language laboratory, (b) using a computer with voice synthesizer and (c) using an advanced student tutor. Following a treatment period, a test is administered to each student with scoring results being pass or fail. The frequency of passing is then recorded for each cell in the 2 by 3 array (gender by treatment). If gender is independent of the treatment variable, the expected frequency of males that pass in each treatment would be the same as the expected frequency for females. The chi-squared statistic is obtained as

$$\chi^2 = \frac{\sum_{i=1}^{row}\sum_{j=1}^{col}(f_{ij}-F_{ij})^2}{F_{ij}} \quad (6.1)$$

where f_{ij} is the observed frequency, F_{ij} the expected frequency, and χ^2 is the chi-squared statistic with degrees of freedom (rows−1) times (columns−1).

Spearman Rank Correlation

When the researcher's data represent ordinal measures such as ranks with some observations being tied for the same rank, the Rank Correlation may be the appropriate statistic to calculate. While the computation for the case of untied cases is the same as that for the Pearson Product–moment correlation, the correction for tied ranks is found only in the Spearman correlation. In addition, the interpretation of the significance of the Rank Correlation may differ from that of the Pearson Correlation where bivariate normalcy is assumed.

Mann–Whitney U Test

An alternative to the Student t-test when the scale of measurement cannot be assumed to be interval or ratio and the distribution of errors is unknown is a non-parametric test known as the Mann–Whitney test. In this test, the dependent variable scores for both groups are ranked and the number of times that one groups scores exceed the rank of scores in the other group are recorded. This total number of times scores in one group exceed those of the other is named U. The sampling distribution of U is known and forms the basis for the hypothesis that the scores come from the same population.

Fisher's Exact Test

The probability of any given pattern of responses in a 2 by 2 table may be calculated from the hypergeometric probability distribution as

$$P = \frac{(A+B)!(C+D)!(A+C)!(B+D)!}{N!A!B!C!D!} \tag{6.2}$$

where A, B, C, and D correspond to the frequencies in the four quadrants of the table and N corresponds to the total number of individuals sampled.

Kendall's Coefficient of Concordance

It is not uncommon that a group of people are asked to judge a group of persons or objects by rank ordering them from highest to lowest. It is then desirable to have some index of the degree to which the various judges agreed, that is, ranked the objects in the same order. The Coefficient of Concordance is a measure varying between 0 and 1 that indicates the degree of agreement among judges. It is defined as:

W = Variance of rank sums / maximum variance of rank sums.

The coefficient W may also be used to obtain the average rank correlation among the judges by the formula:

$$M_r = (mW - 1) / (m - 1) \tag{6.3}$$

where M_r is the average (Spearman) rank correlation, m is the number of judges and W is the Coefficient of Concordance.

Kruskal-Wallis One-Way ANOVA

One-Way, Fixed-Effects Analysis of Variance assumes that error (residual) scores are normally distributed, that subjects are randomly selected from the population and assigned to treatments, and that the error scores are equally distributed in the populations representing the treatments. The scale of measurement for the dependent variable is assumed to be interval or ratio. But what can you do if, in fact, your measure is only ordinal (for example like most classroom tests), and you cannot assume normally distributed, homoscedastic error distributions?

Why, of course, you convert the scores to ranks and ask if the sum of rank scores in each treatment group are the same within sampling error! The Kruskal-Wallis One-Way Analysis of variance converts the dependent score for each subject in the study to a rank from 1 to N. It then examines the ranks attained by subjects in each of the treatment groups. Then a test statistic which is distributed as Chi-Squared with degrees of freedom equal to the number of treatment groups minus one is obtained from:

$$H = \frac{12}{N(N+1)} \sum_{j=1}^{K} R_j^2 / n_j - 3(N+1) \quad (6.4)$$

where N is the total number of subjects in the experiment, n_j is the number of subjects in the jth treatment, K is the number of treatments and R_j is the sum of ranks in the jth treatment.

Wilcoxon Matched-Pairs Signed Ranks Test

This test provides an alternative to the student t-test for matched score data where the assumptions for the parametric t-test cannot be met. In using this test, the difference is obtained between each of N pairs of scores observed on matched objects, for example, the difference between pretest and post-test scores for a group of students. The difference scores obtained are then ranked. The ranks of negative score differences are summed and the ranks of positive score differences are summed. The test statistic T is the smaller of these two sums. Difference scores of 0 are eliminated since a rank cannot be assigned. If the null hypothesis of no difference between the groups of scores is true, the sum of positive ranks should not differ from the sum of negative ranks beyond that expected by chance. Given N ranks, there is a finite number of ways of obtaining a given sum T. There are a total of 2 raised to the N ways of assigning positive and negative differences to N ranks. In a sample of 5 pairs, for example, there are two to the fifth power = 32 ways. Each rank sign would occur with probability of 1/32. The probability of getting a particular total T is

$$P_T = \frac{\text{Ways of getting T}}{2^N} \quad (6.5)$$

The cumulative probabilities for T, T-1,....,0 are obtained for the observed T value and reported. For large samples, a normally distributed z score is approximated and used.

Cochran Q Test

The Cochran Q test is used to test whether or not two or more matched sets of frequencies or proportions, measured on a nominal or ordinal scale, differ significantly among themselves. Typically, observations are dichotomous, that is, scored as 0 or 1 depending on whether or not the subject falls into one or the other criterion group. An example of research for which the Q test may be applied might be the agreement or disagreement to the question "Should abortions be legal?". The research design might call for a sample of n subjects answering the question prior to a debate and following a debate on the topic and subsequently 6 months later. The Q test applied to these data would test whether or not the proportion agreeing was the same under these three time periods. The Q statistic is obtained as

$$Q = \frac{(K-1)\sum_{j=1}^{K} G_j^2 - \left(\sum_{j=1}^{K} G_j\right)^2}{K\sum_{i=1}^{n} L_i - \sum_{i=1}^{n} L_i^2} \tag{6.6}$$

where K is the number of treatments (groups of scores), G_j is the sum with the jth treatment group, and L_i is the sum within case i (across groups). The Q statistic is distributed approximately as Chi-squared with degrees of freedom K-1. If Q exceeds the Chi-Squared value corresponding to the cumulative probability value, the hypothesis of equal proportions for the K groups is rejected.

Sign Test

> Did you hear about the nonparametrician who couln't get his driving license? He couldn't pass the sign test.

Imagine a counseling psychologist who sees, over a period of months, a number of clients with personal problems. Suppose the psychologist routinely contacts each client for a 6 month followup to see how they are doing. The counselor could make an estimate of client "adjustment" before treatment and at the followup time (or better still, have another person independently estimate adjustment at these two time periods). We may assume some underlying continuous "adjustment" variable even though we have no idea about the population distribution of the variable. We are intrested in knowing, of course, whether or not people are better adjusted 6 months after therapy than before. Note that we are only comparing the "before" and "after" state of the individuals with each other, not with other subjects. If we assign a + to the situation of improved adjustment and a—to the situation of same or poorer adjustment, we have the data required for a Sign Test. If treatment has had no effect, we would expect approximately one half the subjects would receive plus signs and the others negative signs. The sampling distribution of the proportion of plus signs is given by the binomial probability distribution with parameter of .5 and the number of events equal to n, the number of pairs of observations.

Friedman Two Way ANOVA

Imagine an experiment using, say, ten groups of subjects with four subjects in each group that have been matched on some relevant variables (or even using the same subjects). The matched subjects in each group are exposed to four different treatments such as teaching methods, dosages of medicine, proportion of positive responses to statements or questions, etc. Assume that some criterion measure on at least a nominal scale is available to measure the effect of each treatment. Now rank the subjects in each group on the basis of their scores on the criterion. We may now ask whether the ranks in each treatment come from the same population. Had we been able to assume an interval or ratio measure and normally distributed errors, we might have used a repeated measures analysis of variance. Failing to meet the parametric test assumptions, we instead examine the sum of ranks obtained under each of the treatment conditions and ask whether they differ significantly. The test statistic is distributed as Chi-squared with degrees of freedom equal to the number of treatments minus one. It is obtained as where N is the number of groups, K the number of treatments (or number of subjects in each group), and R_j is the sum of ranks in each treatment.

Probability of a Binomial Event

Did you hear about the two binomial random variables who talked very quietly because they were discrete?

The BINOMIAL program is a short program to calculate the probability of obtaining k or fewer occurrences of a dichotomous variable out of a total of n observations when the probability of an occurrence is known. For example, assume a test consists of five multiple choice items with each item scored correct or incorrect. Also assume that there are five equally plausible choices for a student with no knowledge concerning any item. In this case, the probability of a student guessing the correct answer to a single item is 1/5 or .20. We may use the binomial program to obtain the probabilities that a student guessing on each item of the test gets a score of 0, 1, 2, 3, 4, or 5 items correct by chance alone.

The formula for the probability of a dichotomous event k where the probability of a single event is p (and the probability of a non-event is $q = 1-p$) is given as:

$$P(k) = \frac{N!}{(N-k)!\,k!}\,p^{(N-k)}\,q^k \qquad (6.7)$$

For example, if a "fair" coin is tossed three times with the probabilities of heads is $p = .5$ (and $q = .5$) then the probabilty of observing 2 heads is

$$P(2) = \frac{3!}{(3-2)!2!}\,0.5^1 \times 0.5^2$$

$$= \frac{3 \times 2 \times 1}{1 \times (2 \times 1)} \times 0.5 \times 0.25$$

$$= \frac{6}{2} \times 0.125 = .375$$

Similarly, the probability of getting one toss turn up heads is

$$P(1) = \frac{3!}{(3-1)!1!}\,0.5^2 \times 0.5 = \frac{6}{2} \times 0.25 \times 0.5 = .375$$

and the probability of getting zero heads turn up in three tosses is

$$P(0) = \frac{3!}{(3-0)!0!}\,0.5^0 \times 0.5^3 = \frac{6}{6} \times 1.0 \times 0.125 = 0.125$$

The probability of getting 2 or fewer heads in three tosses is the sum of the three probabilities, that is, $0.375 + 0.375 + 0.125 = 0.875$.

Runs Test

Random sampling is a major assumption of nearly all statistical tests of hypotheses. The Runs test is one method available for testing whether or not an obtained sample is likely to have been drawn at random. It is based on the order of the values in the sample and the number of values increasing or decreasing in a sequence. For example, if a variable is composed of dichotomous values such as zeros (0) and ones (1) then a run of values such as 0,0,0,0,1,1,1,1 would not likely to have been selected at random. As another example, the values 0,1,0,1,0,1,0,1 show a definite cyclic pattern and also would not likely be found by random sampling. The test involves finding the mean of the values and examining values above and below the mean (excluding values at the mean.) The values falling above or below the mean should occur in a random fashion. A run consists of a series of values above the mean or below the mean. The expected value for the total number of runs is known and is a function of the sample size (N) and the numbers of values above (N1) and below (N2) the mean. This test may be applied to nominal through ratio variable types.

Kendall's Tau and Partial Tau

When two variables are at least ordinal, the tau correlation may be obtained as a measure of the relationship between the two variables. The values of the two variables are ranked. The method involves ordering the values using one of the variables. If the values of the other variable are in the same order, the correlation would be 1.0. If the order is exactly the opposite for this second variable, the correlation would be -1.0 just as if we had used the Pearson Product–moment correlation method. Each pair of ranks for the second variable are compared. If the order (from low to high) is correct for a pair it is assigned a value of $+1$. If the pair is in reverse order, it is assigned a value of -1. These values are summed. If there are N values then we can obtain the number of pairs of scores for one variable as the number of combinations of N things taken 2 at a time which is $N(N-1)$. The tau statistic is the ratio of the sum of 1's and -1's to the total number of pairs. Adjustments are made in the case of tied scores. For samples larger than 10, tau is approximately normally distributed.

Whenever two variables are correlated, the relationship observed may, in part, be due to their common relationship to a third variable. We may be interested in knowing what the relationship is if we partial out this third variable. The Partial Tau provides this. Since the distribution of the partial tau is not known, no test of significance is included.

The Kaplan-Meier Survival Test

Survival analysis is concerned with studying the occurrence of an event such as death or change in a subject or object at various times following the beginning of the study. Survival curves show the percentage of subjects surviving at various times as the study progresses. In many cases, it is desired to compare survival of an experimental treatment with a control treatment. This method is heavily used in medical research but is not restricted to that field. For example, one might compare the rate of college failure among students in an experimental versus a control group.

Kolmogorov-Smirnov Test

One often is interested in comparing a distribution of observed values with a theoretical distribution of values. Because many statistical tests assume a "normal" distribution, a variety of tests have been developed to determine whether or not two distributions are different beyond that expected due to random sampling variations. This test lets you compare your distribution with several theoretical distributions.

Chapter 7
Statistical Process Control

Introduction

Statistical Process Control (SPC) has become a major factor in the reduction of manufacturing process errors over the past years. Sometimes known as the Demming methods for the person that introduced them to Japan and then the United States, they have become necessary tools in quality control processes. Since many of the employees in the manufacturing area have limited background in statistics, a large dependency has been built on the creation of charts and their interpretation. The statistics which underlay these charts are often those we have introduced in previous sections. The unique aspect of SPC is in the presentation of data in the charts themselves.

XBAR Chart

In quality control, observations are typically made in "lots", that is, a number of observations are made on some product's manufacturing process or the product itself at periodic intervals. For example, in the manufacture of metal bolts, the length of bolts being turned out may be sampled each hour of the day. The means and standard deviation of these sample lots may then be calculated and plotted with lines drawn to show the overall mean and upper and lower "control limits" indicating whether or not a process may be "out of control". One area of confusion which exists is the language used by industrial people in indicating their level of process control. You may hear the expression that "we employ control to 6 sigmas." They **do not mean** they use 6 standard deviations as their upper and lower control limits but rather that the probability of being out of control is that associated with the normal curve probability of a value being 6 standard deviations or greater (a very small value.) This confusion of standard deviations (sigmas) and the probability associated with departures from the mean under the normal distribution assumption is unfortunate. When you select the sigma values for control limits, the limits for 1 sigma are much closer to the mean that for 3 sigma. You may, of course, select your own limits that you feel are practical for your process control. Since variation in raw materials, tool wear, shut-down costs for replacement of worn tool parts, etc. may be beyond your control, limits must be set that maximize quality and minimize costs.

Range Chart

As tools wear the products produced may begin to vary more and more widely around the values specified for them. The mean of a sample may still be close to the specified value but the range of values observed may increase. The result is that more and more parts produced may be under or over the specified value. Therefore quality assurance personnel examine not only the mean (XBAR chart) but also the range of values in their sample lots.

S Control Chart

The sample standard deviation, like the range, is also an indicator of how much values vary in a sample. While the range reflects the difference between largest and smallest values in a sample, the standard deviation reflects the square root of the average squared distance around the mean of the values. We desire to reduce this variability in our processes so as to produce products as similar to one another as is possible. The S control chart plot the standard deviations of our sample lots and allows us to see the impact of adjustments and improvements in our manufacturing processes.

CUSUM Chart

The cumulative sum chart, unlike the previously discussed SPC charts (Shewart charts) reflects the results of all of the samples rather than single sample values. It plots the cumulative sum of deviations from the mean or nominal specified value. If a process is going out of control, the sum will progressively go more positive or negative across the samples. If there are M samples, the cumulative sum S is given as

$$S = \sum_{i=1}^{M} (\overline{X}_i - \mu_o)$$

where $\overline{X}_i$ is the observed sample mean and μ_o is the nominal value or (overall mean.)

It is often desirable to draw some boundaries to indicate when a process is out of control. By convention we use a standardized difference to specify this value. For example with the boltsize.txt data, we might specify that we wish to be sensitive to a difference of 0.02 from the mean. To standardize this value we obtain

$$\delta = \frac{0.02}{\sigma_X}$$

or using our sample values as estimates obtain

$$\delta = \frac{0.02}{S_X} = \frac{0.02}{0.359} = 0.0557$$

A "V Mask" is then drawn starting at a distance "d" from the last cumulative sum value with an angle θ back toward the first sample deviation. In order to calculate the distance d we need to know the probabilities of a Type I and Type II error, that is, the probability of incorrectly concluding that a shift to out-of-control has taken place and the probability of failing to detect an out-of-control condition. If these values are specified then we can obtain the distance d as

$$d = \left(\frac{2}{\delta^2}\right) \ln\left(\frac{1-\beta}{\alpha}\right)$$

When you run the CUSUM procedure you will note that the alpha and beta error rates have been set to default values of 0.05 and 0.20. This would imply that an error of the first type (concluding out-of-control when in fact it is not) is a more "expense" error than concluding that the process is in control when in fact it is not. Depending on the cost of shut-down and correction of the process versus scraping of parts out of tolerance, you may wish to adjust these default values.

The angle of the V mask is obtained by

$$\theta = \tan^{-1}\left(\frac{\alpha}{2K}\right)$$

where k is a scaling factor typically obtained as $k = 2\sigma_x$

The specification form for the CUSUM chart is shown below for the data file labeled boltsize.txt. We have specified our desire to detect shifts of 0.02 in the process and are using the 0.05 and 0.20 probabilities for the two types of errors.

p Chart

In some quality control processes the measure is a binomial variable indicating the presence or absence of a defect in the product. In an automated production environment, there may be continuous measurement of the product and a "tagging" of the product which is non-conforming to specifications. Due to variation in materials, tool wear, personnel operations, etc. one may expect that a certain proportion of the products will have defects. The p Chart plots the proportion of defects in samples of the same size and indicates by means of upper and lower control limits, those samples which may indicate a problem in the process.

Defect (Non-Conformity) c Chart

The previous section discusses the proportion of defects in samples (p Chart.) This section examines another defect process in which there is a count of defects in a sample lot. In this chart it is assumed that the occurrence of defects are independent, that is, the occurrence of a defect in one lot is unrelated to the occurrence in another lot. It is expected that the count of defects is quite small compared to the total number of parts potentially defective. For example, in the production of light bulbs, it is expected that in a sample of 1000 bulbs, only a few would be defective. The underlying assumed distribution model for the count chart is the Poisson distribution where the mean and variance of the counts are equal.

Defects Per Unit u Chart

Like the count of defects c Chart described in the previous section, the u Chart describes the number of defects per unit. It is assumed that the number of units observed is the same for all samples. We will use the file labeled uChart.txt as our example. In this set of data, 25 observations of defects for 45 units each are recorded. The assumption is that defects are distributed as a Poisson distribution with the mean given as

$\bar{u} = \dfrac{\sum c}{\sum n}$ where c is the count of defects and n is the number of units observed. and

$$\text{UCL} = \bar{u} + \text{sigma}\sqrt{\dfrac{\bar{u}}{n}} \text{ and } \text{LCL} = \bar{u} - \text{sigma}\sqrt{\dfrac{\bar{u}}{n}}$$

Chapter 8
Linear Programming

Introduction

Linear programming is a subset of a larger area of application called mathematical programming. The purpose of this area is to provide a means by which a person may find an optimal solution for a problem involving objects or processes with fixed 'costs' (e.g. money, time, resources) and one or more 'constraints' imposed on the objects. As an example, consider the situation where a manufacturer wishes to produce 100 lb of an alloy which is 83% lead, 14% iron and 3% antimony. Assume he has at his disposal, five existing alloys with the following characteristics:

Alloy1	Alloy2	Alloy3	Alloy4	Alloy5	Characteristic
90	80	95	70	30	Lead
5	5	2	30	70	Iron
5	15	3	0	0	Antimony
$6.13	$7.12	$5.85	$4.57	$3.96	Cost

This problem results in the following system of equations:

$$X_1 + X_2 + X_3 + X_4 + X_5 = 100$$
$$0.90X_1 + 0.80X_2 + 0.95X_3 + 0.70X_4 + 0.30X_5 = 83$$
$$0.05X_1 + 0.05X_2 + 0.02X_3 + 0.30X_4 + 0.70X_5 = 14$$
$$0.05X_1 + 0.15X_2 + 0.03X_3 = 3$$
$$6.13X_1 + 7.12X_2 + 5.85X_3 + 4.57X_4 + 3.96X_5 = Z(min)$$

The last equation is known as the 'objective' equation. The first four are constraints. We wish to obtain the coefficients of the X objects that will provide the minimal costs and result in the desired composition of metals. We could try various combinations of the alloys to obtain the desired mixture and then calculate the price of the resulting alloy but this could take a very long time!

As another example: a dietitian is preparing a mixed diet consisting of three ingredients, food A, B and C. Food A contains 81.85 g of protein and 13.61 g of fat and costs 30 cents per unit. Each unit of food B contains 58.97 g of protein and 13.61 g of fat and costs 40 cents per unit. Food C contains 68.04 g of protein and 4.54 g of fat and costs 50 cents per unit. The diet being prepared must contain the at least 100 g of protein and at the most 20 g of fat. Also, because food C contains a compound that is important for the taste of the diet, there must be exactly 0.5 units of food C in the mix. Because food A contains a vitamin that needs to be included, there should also be a minimum of 0.1 units of food A in the diet. Food B contains a compound that may be poisonous when taken in large quantities, and the diet may contain a maximum of 0.7 units of food B. How many units of each food should be used in the diet so that all of the minimal requirements are satisfied, the maximum allowances are not violated, and we have a diet which cost is minimal? To make the problem a little bit easier, we put all the information of the problem in a tableau, which makes the formulation easier.

	Protein	Fat	Cost	Minimum	Maximum	Equal
Food A	81.65	13.60	$0.30	0.10		
Food B	58.97	13.60	$0.40		0.70	
Food C	68.04	4.54	$0.50			0.5
Min.	100					
Max.		20				

The numbers in the tableau represent the number of grams of either protein and fat contained in each unit of food. For example, the 13.61 at the intersection of the row labelled "Food A" and the column labelled "Fat" means that each unit of food A contains 13.61 g of fat.

Calculation

We must include 0.5 units of food C, which means that we include $0.5 * 4.54 = 2.27$ g of fat and $0.5 * 68.04 = 34.02$ g of protein in the diet, coming from food C. This means, that we have to get $100 - 34.02 = 65.98$ g of protein or more from Food A and B, and that we may include a maximum of $20 - 2.27 = 17.73$ g of fat from food A and B. We have to include a minimum of 0.1 units of food A in the diet, accounting for 8.17 g of protein and 0.45 g of fat. This means that we still have to include $65.98 - 8.17 = 57.81$ g of protein from food A and/or B, and that the maximum allowance for fat from A and/or B is now $17.73 - 0.45 = 17.28$ g. We should first look at the cheapest possibility, e.g. inclusion of food A for the extra required 57.81 g of protein. If we include $57.81/81.65 = 0.708$ units of food A, we have met the requirement for protein, and we have added $0.708 * 13.61 = 9.64$ g of fat, which is below the allowance of 17.28 g which had remained. So we don't need any of the food B, which is more expensive, and which is contains less protein. The price of the diet is now $0.48. But what would we do, if food B was available at a lower price? We may or may not want to use B as an ingredient. The more interesting question is, at what price would it be interesting to use B as an ingredient instead of A? This could be approached by an iterative procedure, by choosing a low price for B, and see if the price for the diet would become less than the calculated price of $0.48.

Implementation in Simplex

A more sophisticated approach to these problems would be to use the Simplex method to solve the linear program. The sub-program 'Linear Programming', provided with OpenStat can be used to enter the parameters for these problems in order to solve them.

Chapter 9
Measurement

Three roommates slept through their midterm statistics exam on Monday morning. Since they had returned together by car from the same hometown late Sunday evening, they decided on a great little falsehood. The three met with the instructor Monday afternoon and told him that an ill-timed flat tire had delayed their arrival until noon. The instructor, while somewhat skeptical, agreed to give them a makeup exam on Tuesday. When they arrived the instructor issued them the same makeup exam and ushered each to a different classroom. The first student sat down and noticed immediately the instructions indicated that the exam would be divided into Parts I and II weighted 10% and 90% respectively. Thinking nothing of this disparity, he proceeded to answer the questions in Part I. These he found rather easy and moved confidently to Part II on the next page. Suddenly his eyes grew large and his face paled. Part II consisted of one short and pointed question....... "Which tire was it?"

Evaluators base their evaluations on information. This information comes from a number of sources such as financial records, production cost estimates, sales records, state legal code books, etc. Frequently the evaluator must collect additional data using instruments that he or she alone has developed or acquired from external sources. This is often the case for the evaluation of training and educational programs, evaluation of personnel policies and their impacts, evaluation of social and psychological environments of the workplace, and the evaluation of proposed changes in the way people do business or work.

This chapter will give guidance in the development of instruments for making observations in the cognitive and affective domains of human behavior.

Test Theory

The sections presented below provide a detailed discussion of testing theory. You do not need to understand all of this theory to make appropriate use of tests in your evaluations, although it may help in avoiding some errors in decisions or selecting appropriate analytic tools. It is included for the "advanced" student of evaluation who is responsible as an "expert" in assisting other evaluators in correctly using and analyzing tests. If you are "afraid" of statistics, you may skip the formal "proofs" of the equations and focus primarily on the resulting equations.

Theory and practice are the same in theory. In practice they are different.

Scales of Measurement

Measurement is the assignment of a label or number to an object or person to characterize that individual on the basis of an observed attribute. The manner in which we make our observations will determine our "scale of measurement."

Nominal Scales

Sometimes we observe an attribute in such a way that we can only classify an individual or object as possessing or not possessing the attribute. For example, the variable "gender" may be observed in such a way as to permit only labeling an individual as "male" or not male (female). The attribute of "country of origin" may lead us to classify individuals by their place of birth such as "USA", "Canada", "European", etc. The assignment of labels or names to objects based on a specific attribute is called a NOMINAL scale of measurement. We can, of course, arbitrarily select the labels to assign the observed individuals. Letters such as "A", "B", "C", etc. might be used or even numbers such as "1", "2", "3", etc. Notice, however that the use of numbers as labels may cause some confusion with the use of numbers to indicate a quantity of some attribute. When using a nominal scale of measurement, there is no attempt to indicate quantity. Coding males as 1 and females as 0, for example, would not indicate males are "greater" on some quantitative variable—we might just as well have assigned 1 to females and 0 to males!

Ordinal Scales of Measurement

Some attributes of individuals or objects may be observed in such a way that the individuals may be ordered, that is, arranged in a manner that indicates person "B" possesses more of the attribute than person "A", but less than person "C". For example, the number of correctly answered items on a test may permit us to say that John has a higher score than Mary but a lower score than Jim. (NOTE! We carefully avoided saying that John knows more than Mary but knows less than Jim. Such statements imply a direct relationship between the amount of knowledge of a subject and the number of items passed. This is virtually never the case!) When we assign numbers that only indicate the ordering of individuals on some attribute, our scale of measurement is called an ordinal scale. We will add that comparing the means of groups measured with an ordinal scale leads to serious problems of interpretation. The median, on the other hand, is more interpretable.

Interval Scales of Measurement

There is a class of measurements known as interval scales of measurement. These refer to observing an attribute of individuals in such a way that the numbers assigned to individuals denote the relative amount of the attribute possessed by that individual in comparison to some "standard" or referent. The assignment of numbers in this way would permit a transformation (such as multiplying all numbers by a constant) that would preserve the proportional distance among the individuals. The numbers assigned do not indicate the absolute amount of the attribute—only the amount relative to the standard. For example, we might say that the average number of questions answered correctly on a test of 100 items measuring recall of nonsense words by a very large population of 18 year old males constitutes our "standard". IF all items are equally difficult to recall, we might use the proportion of the standard number of items recalled as an interval measure of recall ability. That is, the difference between Mary who obtains a score of 20 and John who receives a score of 40 is proportional to the difference between John and Jim who receives a score of 60. Even if we multiply their scores by 100, the distance between Mary, John and Jim is proportionally the same! Again note that the proportion of the standard number of items correctly recalled is NOT a measure of individual's ability to recall items in general. It is only their ability to recall the carefully selected items of this test in comparison to the standard that is measured. A different set of items could lead to assignment of a completely different set of numbers to each individual with different relative distances among the individuals. As another example, consider a measure of individual "wealth". Assume wealth is defined as the total of a person's debts and credits using the standard "dollar". We may clearly have individuals with negative "wealth" (debts exceed credits) and individuals with "positive" wealth (credits exceed debts). Our wealth scale has equal intervals (dollars). We can make statements such as John has 20 dollars more wealth than Mary but five dollars less wealth than Jim. In other words, we can represent the distance among our individuals as well as their order. Note, however, that an individual with a wealth score of zero (debts = assets) is NOT broke, that is, have an absence of wealth. With an interval scale of measurement 0.0 does NOT mean an absence of the attribute—only a relative amount compared to the "standard". Zero is an arbitrary point on our scale of measurement:

```
                Personal Wealth
    "Mary"         "John"          "JIM"
    ─────────────────────────────────────
     -10            0      +10     +15
                 dollars
```

If a test of, say, 20 history items consists of items that are equally increasing in difficulty, we may use such a test to indicate the distance among subjects administered the test. We do, however, require that if an individual misses an item with known difficulty dj, that the same individual will miss all items of greater difficulty! Please note that missing all items does not mean an absence of knowledge! (We might have included easier items.) We may also have assigned "scores" to our subjects as $X =$ the number of items "passed"—the number of items "missed". Again, the zero point on our scale is arbitrary and does not reflect an actual amount of knowledge or absence of knowledge! Tests of intelligence, achievement or aptitude may be constructed that utilize an interval scale of measurement. Like the value of a "dollar", the "difficulty" of each item must be clearly defined. We can say, for example, that $100.00 buys an ounce of silver. We might similarly define an item of difficulty 1.0 as that item which is correctly answered by 50 % of 18 year old male freshmen college students residing in the USA in 1988.

Ratio Scales of Measurement

We may sometimes observe an attribute of an individual or object in such a way that the numeric values assigned the individuals indicate the actual amount of the attribute.

For example, we might measure the time delay between the occurrence of a stimulus (e.g. the flash of a strobe light) and the observation on the surface of the brain of a change in electrical potential representing response to the stimulus. Such an observed latency may theoretically vary from 0 to infinity in whatever units of time (e.g. microseconds) that we wish to utilize. We could then make statements such as John's latency is twice as long as Mary's latency but half as long as Jim's latency. Note that a zero latency is meaningful and not an arbitrary point on the scale! Another example of a ratio scale of measurement is the distance, perhaps in inches, that a person can jump. In each case, the ratio scale of measurement has a "true" zero point on the scale which can be interpreted as an absence of the attribute. In addition, the ratio scale permits forming meaningful ratios of subject's scores. For example we might say that John can jump twice as far as Mary but Jim (who is in a wheelchair) cannot jump at all! Could we ever construct a test of intelligence that yielded ratio scale numbers? What would a statement that Mary is twice as intelligent as John but half as intelligent as Jim mean? What would a score of zero intelligence mean? What would a score of 1.0 mean? Clearly, it is difficult, if not nearly impossible to construct ratio scale measures for attributes that we cannot directly observe and for which we have no meaningful "standard" with which to relate. We may, in fact, be hard-pressed to provide evidence that our psychological and educational measurement scales are even interval scales. Many are clearly only ordinal measures at best.

Reliability, Validity and Precision of Measurement

Reliability

If we stepped on and off our weight scale and each time received a different reading for our weight, we would probably go out and buy a new scale! We would say we want a reliable scale—one that consistently yields the same weight for the same object measured. When we refer to tests, the ability of a test to produce the same values when used to measure the same subjects is also called the reliability of the test. If we carefully examine the "markings" on our weight scale however, we might be surprised that there are, in fact, some variations in the values we could record. Sometimes I might weight 150.3 and the next time I get on the scale I observe 150.2. Did the scale actually give different values or was I only able to interpret the distance between the marks for 150 and 151 approximately and therefore introduce some "error" or variation in the values recorded? This lack of sufficient "in-between" markings on our scale is referred to as the precision of our measurement. If the scale is only marked in whole pounds, my precision of observation is limited to whole pounds. In fact, when the scale appears right in between 150 and 151, is the closest value 150 or 151? My error of precision is potentially 1 lb. Note that precision is NOT the same as reliability. When we speak of reliability, we are speaking of variations in repeated observations that are larger than those due to the precision of measurement alone.

In describing the reliability of an instrument, it is advantageous to have an index which describes the degree of reliability of the instrument. One popular index of reliability is the product–moment correlation between two applications of the measurement instrument to a group of individuals. For example, I might administer a history test to a group of students at 10:00 a.m. and again at 2:00 p.m. Assuming the students did not talk with each other about the test, study history during the intervening time, forget relevant history material during those 4 h, etc., then the correlation between their 10:00 a.m. and 2:00 p.m. scores would estimate the reliability of the test. Our index of reliability can vary between zero (no reliability) to 1.0 (perfect reliability). Note that a reliability of less than zero is nonsense—a test cannot theoretically be less than completely unreliable!

We may also express this index of reliability as the ratio of "True Score" variance to "Observed Score Variance", that is S_t^2/S_x^2. We will denote this ratio as r_{xx}. This choice of r_{xx} is not capricious—we use the symbol for correlation to indicate that reliability is estimated by a product–moment correlation coefficient.

The xx subscript denotes a correlation of a measure with itself. Each observed score (X) for an individual may be assumed to consist of two parts, a TRUE score (T) and an ERROR (E) score, i.e., $X_i = T_i + E_i$. For N individuals, the variance of the observed scores is

$$S_x^2 = \frac{\sum_{i=1}^{N}[(T_i + E_i) - \overline{(T_i + E_i)}]^2}{(N-1)}$$

or

$$S_x^2 = \frac{\sum_{i=1}^{N}[T_i + E_i - \bar{T} - \bar{E}]^2}{(N-1)}$$

or

$$S_x^2 = \frac{\sum_{i=1}^{N}[(T_i - \bar{T}_i) + (E_i - \bar{E})]^2}{(N-1)} \quad (9.1)$$

If we assume that error scores (E) are normally and randomly distributed with a mean of zero and, since they are random, uncorrelated with other scores, then

$$S_x^2 = \frac{\sum_{i=1}^{N}[(T_i - \bar{T}) + E_i]^2}{(N-1)}$$

$$= \frac{\sum_{i=1}^{N}[(T_i - \bar{T}) + E_i^2 + (E_i T_i - E_i \bar{T})]^2}{(N-1)}$$

$$= \frac{\sum_{i=1}^{N}(T_i - \bar{T})^2}{(N-1)} + \frac{\sum_{i=1}^{N} E_i^2}{(N-1)} + \frac{\sum_{i=1}^{N} E_i(T_i - \bar{T})}{(N-1)}$$

$$= S_t^2 + S_e^2 + \text{Cov}_{te}/(N-1)$$
$$= S_t^2 + S_e^2 + \text{Cov}_{te}/(N-1) * (S_t S_e)/(S_t S_e)$$
$$= S_t^2 + S_e^2 + r_{te} S_t S_e$$
$$= S_t^2 + S_e^2 \text{ since the correlation of errors with true scores is zero.} \quad (9.2)$$

Reliability is defined as

$$r_{xx} = \frac{S_t^2}{S_x^2} = \frac{S_x^2 - S_e^2}{S_x^2} = \frac{1 - S_e^2}{S_x^2} \quad (9.3)$$

Because we cannot directly observe true scores, we must estimate them (or the variance of error scores) by some method. A variety of methods have been developed to estimate the reliability of a test. We will describe, in this unit, the one known as the Kuder-Richardson Formula 20 estimate. Other methods include the test-retest method, the corrected split-half method, the Cronbach Alpha method, etc.

The Kuder: Richardson Formula 20 Reliability

The K-R formula is based on the correlation between a test composed of K observed items and a theoretical (unobserved) parallel test of k items parallel to those of the observed test. A parallel test or item is one which yields the same means, standard deviations and intercorrelations as the original ones.

To develop the K-R 20 formula, we will begin with the correlation between two tests composed of K and k items respectively where $K = k$. The correlation between the total scores correct on each test is represented by

$r_{I,II}$ where
Test I scores = the sum of item scores $X_1 + X_2 + .. + X_K$
and Test II scores = the sum of item scores $x_1 = x_2 + .. + x_k$

We may therefore write the correlation as

$$r_{I,II} = r_{(X_1+X_2+..+X_K),(x_1+x_2+..+x_k)}$$

$$= \frac{\sum_{i=1}^{N} [(X_1 + .. + X_K) - (\overline{X_1 + .. + X_K})][(x_1 + .. + x_k) - (\overline{x_1 + ..x_k})]}{N \sum_{G=1}^{K} S_g^2 + N \sum_{G=1}^{K}\sum_{g=1}^{K} r_{G,g}S_G S_g} \quad g \neq G$$

(9.4)

The numerator of the above equation is the deviation cross-products of the total scores I and II. The denominator represents the variance of the composite score I. Since parallel tests have the same variance, we are assuming that the variance of test I equals that of test II. For that reason, the variance of the composite test I or II can be expressed as the sum of individual item variances plus the covariance among the items. The numerator of our correlation can be similarly expressed, that is

$$r_{I,II} = \frac{N \sum_{G=1}^{K} r_{G,G}S_G S_g + N \sum_{G=1}^{K}\sum_{g=1}^{K} r_{g,G}S_G S_g}{N \sum_{g=1}^{K} S_g^2 + N \sum_{g=1}^{K}\sum_{G=1}^{K} r_{g,G}S_g S_G} \quad g \neq G$$

which can be further reduce as follows:

$$r_{I,II} = \frac{\sum_{g=1}^{K} r_{g,g}S_g^2 + \sum_{g=1}^{K}\sum_{G=1}^{K} r_{g,G}S_gS_G}{\sum_{g=1}^{K} S_g^2 + \sum_{g=1}^{K}\sum_{G=1}^{K} r_{g,G}S_gS_G} \qquad g \neq G$$

$$r_{I,II} = \frac{\sum_{g=1}^{K} r_{g,g}S_g^2 - \sum_{g=1}^{K} S_g^2 + \sum_{g=1}^{K} S_g^2 + \sum_{g=1}^{K}\sum_{G=1}^{K} r_{g,G}S_gS_G}{S_x^2}$$

$$r_{I,II} = \frac{\sum_{g=1}^{K} r_{g,g}S_g^2 - \sum_{g=1}^{K} S_g^2 + S_x^2}{S_x^2} \tag{9.5}$$

Note! $r_{g,g}$ represents the correlation between parallel test items.

In an observed test of K items we would not expect to have parallel items. We must therefore estimate the correlation (or covariance) among parallel items by the correlation among non-parallel items. That is

$$\sum_{g=1}^{K} r_{g,g}S_g^2 = \frac{\sum_{g=1}^{K}\sum_{G=1}^{K} r_{g,G}S_gS_G}{(K-1)} \qquad g \neq G$$

Note: There are K(K−1) pairings when g is not equal to G.
Since

$$S_x^2 = \sum_{g=1}^{K} S_g^2 + \sum_{g=1}^{K}\sum_{G=1}^{K} r_{g,G}S_gS_G$$

then

$$\sum_{g=1}^{K} r_{g,g}S_g = \left(S_x^2 - \sum_{g=1}^{K} S_g^2\right)/(K-1)$$

and

$$r_{I,II} = \frac{\frac{S_x^2 - \sum_{g=1}^{K} S_g^2}{K-1} - \sum_{g=1}^{K} S_g^2 + S_x^2}{S_x^2}$$

$$= \frac{S_x^2 - \sum_{g=1}^{K} S_g^2}{(K-1)S_x^2} - \frac{\sum_{g=1}^{K} S_g^2}{S_x^2} + 1$$

$$= \frac{1}{K-1} \frac{\sum_{g=1}^{K} S_g^2}{S_x^2} \frac{1}{K-1} - \frac{(K-1)\sum_{g=1}^{K} S_g^2}{(K-1)S_x^2} + \frac{K-1}{K-1}$$

$$= \frac{[1]}{K-1} \left[1 - \frac{\sum_{g=1}^{K} S_g^2}{S_x^2} - \frac{K \sum_{g=1}^{K} S_g^2}{S_x^2} + \frac{\sum_{g=1}^{K} S_g^2}{S_x^2} + K - 1 \right]$$

$$= \frac{1}{K-1} \left[K - \frac{K \sum_{g=1}^{K} S_g^2}{S_x^2} \right]$$

$$\text{or } r_{I,II} = \left[\frac{K}{K-1} \right] \left[1 - \sum_{g=1}^{K} S_g^2 / S_x^2 \right] \qquad \text{KR\#20 Formula} \qquad (9.6)$$

We have thus derived the Kuder-Richardson Formula 20 estimate of the correlation between an observed test of K items and a theoretically parallel test of k items. Besides knowing the number of items K, one must calculate the sum of the item variances for item $g = 1$ to K and the total variance of the test (S_x^2). We really only had to make one assumption other than the parallel test assumptions: that the covariance among UNLIKE items is a reasonable estimate of covariance among PARALLEL items.

If we might also assume that all items are equally difficult (they would have the same means and variances) then the above formula may be even further simplified to

$$r_{xx} = \frac{K}{K-1} \left[1 - \frac{\bar{X} - \bar{X}^2 / K}{S_x^2} \right] \qquad (9.7)$$

We note that in the KR#20 formula, that as the number of items K grows large, the ratio of K/(K−1) approaches 1.0 and the reliability approaches

$$r_{xx} = \frac{S_x^2 - \sum_{g=1}^{K} S_g^2}{S_x^2}$$
$$= S_t^2 / S_x^2$$

We now have an expression for the variance of true scores, that is $S_t^2 = S_x^2 r_{xx}$. Similarly, we may obtain an expression for the variance of errors by

$$r_{xx} = \left(S_x^2 - S_e^2\right) / S_x^2$$
$$= 1.0 - S_e^2 / S_x^2$$

$$\text{or} \quad S_e^2 = S_x^2 (1 - r_{xx}) \quad (9.8)$$

The Standard Error of Measurement, the positive root of the variance of errors is obtained as

$$S_e = S_x \sqrt{(1.0 - r_{xx})} \quad (9.9)$$

If the errors of measurement may be assumed to be normally distributed, the standard error indicates the amount of score variability to be expected with repeated measures of the same object. For example, a test that has a standard deviation of 15 and a reliability of 0.91 (as estimated by the KR#20 formula) would have a standard error of measurement of $15 * 0.3 = 4.5$. Since one standard deviation of the normal curve encompasses approximately 68.2 % of the scores, we may say that approximately 68 % of an individual's repeated measurements would be expected to fall within + or −4.5 raw score points. We take note of the fact that this is the error of measurement expected of all individuals measured by a hypothetical instrument no matter what the original score level observed is. If you read about the Rasch method of test analysis, you will find that there are different estimates of measurement error for subjects with varying score levels by that method!

Validity

When we develop an instrument to observe some attribute of objects or persons, we assume the resulting scores will, in fact, relate to that attribute. Unfortunately, this is not always the case. For example, a teacher might construct a paper and pencil test of mathematics knowledge. If a student is unable to read (perhaps blind) then the test would not be valid for that individual. In addition, if the teacher included many "word" problems, the test scores obtained for students may actually measure reading ability to a greater extent than mathematics ability! The "ideal" measurement instrument yields scores indicative of only the amount (or relative amount compared with others) of the single attribute of a subject. It is NOT a score reflecting multiple attributes.

Consider, for a moment, that whenever you wanted a measure of someone's weight, your scale gave you a combination of both their height and weight! How would you differentiate among the short fat persons and the tall thin persons since they could have identical scores? If a test score reflects both mathematics and reading ability, you cannot differentiate persons good in math but poor in reading from those poor in math but good in reading!

The degree to which a test measures what it is intended to measure is called the VALIDITY of the test. Like reliability, we may use an index that varies between 0 and 1.0 to indicate the validity of a test. Again, the Pearson product–moment correlation coefficient is the basis of the validity index.

Concurrent Validity

If there exists another test in which we have confidence of it being reasonable measure of the same attribute measured by our test, we may use the p-m correlation between our test and this "criterion" test as a measure of validity. For example, assume you are constructing a new test to measure the aptitude that students have for learning a foreign language. You might administer your test and the Modern Foreign Language Aptitude Test to the same group of subjects. The correlation between the two tests would be the validity coefficient.

Predictive Validity

Some tests are intended to be used as predictors of some future attribute. For example, the Scholastic Aptitude Test (SAT) may be useful as a predictor of future Grade Point Average earned by students in their freshman year at college. When we correlate the results of a test administered at one point in time with a criterion measured at some future time, the correlation is a measure of the predictive validity of the test.

Discriminate Validity

Some tests which purportedly measure a single attribute are, as we have said, often composite measures of multiple attributes. Ideally, an English test would correlate highly with other English tests and NOT particularly high with intelligence tests, mathematics tests, mechanical aptitude tests, etc. The degree to which the correlation with similar attribute measures differs from the correlation of our test with measures of other attributes is called the discriminate validity of a test. Often the partial correlation between two tests in which the effects of a third, supposedly less related test, has been removed, is utilized as a discriminate validity coefficient. As an example, assume that your new test of English correlates 0.8 with student final examination scores in an English course and correlates 0.5 with the Stanford-Binet test of intelligence. Also assume that the final examination scores correlate 0.4 with the S-B IQ scores. The partial correlation of your English Test with English final examination scores can be obtained as

$$r_{y,E.I} = \frac{r_{y,E} - r_{y,I} r_{E,y}}{\sqrt{(1 - r_{y,I}^2)(1 - r_{e,y}^2)}} \qquad (9.10)$$

where

$r_{y,E.I}$ is the partial correlation between your test y and the English examination scores,
$r_{y,E}$ is the correlation of your test and the English examination scores,
$r_{y,I}$ is the correlation of your test with IQ scores, and
$r_{E,y}$ is the correlation between English examination scores and IQ scores.

The obtained value would be

$$\begin{aligned} r_{y,E.I} &= [(.8 - (.5)(.4)] / \sqrt{[(1 - .25)(1 - .16)]} \\ &= .6 / \sqrt{[(.75)(.84)]} \\ &= .6 / \sqrt{(.63)} = .75 \end{aligned}$$

In other words, partialling out the effects of intelligence reduced our validity from 0.8 to 0.75.

It is sometimes distressing to discover that a carefully constructed test of a single attribute often may be found to correlate substantially with a number of other tests which supposedly measure other, unrelated attributes. In our example, we partial out only the effects of one other variable, intelligence. One can use multiple regression procedures to partial out more than one variable from a correlation.

Construct Validity

The attribute we are proposing to measure with a test is often simply a hypothetical construct, that is, some attribute we think exists but which we have had to define by simple description in our language. There is often no way to directly observe the attribute. The concept of "intelligence" is such a hypothetical construct. We describe more "intelligent" people as those who learn faster and retain their learning longer. Less "intelligent" persons seem to learn at a much slower pace and have more difficult time retaining what they have learned. With such descriptions, we may construct an "intelligence" test. As you probably well know, a number of people have, in fact, done just that! Now assume that your "intelligence" test along with that of, say, three other tests of intelligence, are all administered to the same group of subjects. We could then construct the inter-correlation matrix among these four tests and ask "is there one common underlying variable that accounts for the major portion of variance and covariance within and among these tests?" This question is often answered by determining the eigenvalues and corresponding eigenvectors of the correlation matrix. If there is one particularly larger root out of the four possible roots and if the normalized corresponding eigenvalues of that root all are large, we may argue that there is validity for the construct of intelligence (at least as defined by the four tests). This technique and others similar to it are usually called "Factor Analysis." If our test "loads" (correlates) highly with the same common factor that the other tests measuring the same attribute do, then we argue the test has construct validity. This correlation (factor loading) of our test with the other measures of the same attribute is the construct validity coefficient of our test.

Content Validity

If you were to construct a test of knowledge in a specific area, say "proficiency in statistics", then the items you elect to include in your test should stand the scrutiny of experts in the field of statistics. That is, the content of your test in terms of the items you have written should be relevant to the attribute to be measured. When constructing a test, an initial decision is made as to the purpose of the test: is the purpose to demonstrate proficiency to some specified level, or is it to measure the degree of knowledge attained as compared to others. The first type of test is often referred to as a "criterion" referenced test. The second type in a normative test. With a criterion referenced test, the test writer is usually not as concerned with measuring a "single" attribute or latent variable but rather of selecting items that demonstrate specific knowledge and skills required for doing a certain job or success in some future learning activity. The norm-referenced tests, on the other hand, usually measure the degree of some predominant attribute or "latent" (underlying) variable. In either case, the test author will typically start with a "blueprint" of the domain, i.e., a list of the relevant aspects of the attribute to be measured. This blueprint may be a two-dimensional description of both the topics included in the domain as well as the levels of complexity or difficulty to be measured by items within one aspect. Once the blueprint is constructed, it is used to guide the construction of items so that the domain is adequately sampled and represented by the test. When completed, the test may be submitted to a panel of experts who are asked to classify the items into the original blueprint, evaluate the relevance of the blueprint areas and items constructed and evaluate the adequacy of the item construction. The percent of agreement among judges on a particular item as being appropriate or not being appropriate as a measure of the attribute can be used as an indicator of content validity. The reliability of judgments across a set of items may be used to measure the consistency of the judges themselves. A large proportion of the test items should be judged satisfactory by a high percentage of the judges in order to say that the instrument has content validity.

Effects of Test Length

Tests of achievement, aptitude, and ability may vary considerably in their number of items, i.e. test length. Tests composed on positively correlated items that are longer will display higher reliability then shorter tests. The correlation of reliable measures with other variables will tend to be higher than the correlation of less reliable measurements, thus the predictive validity, concurrent validity, etc. will be higher for the longer test.

Reliability for tests that have been changed in length by a factor of K can be estimated by the Spearman-Brown "prophecy" formula:

$$R_{kk} = \frac{K \, r_{11}}{1 + (K - 1) \, r_{11}} \quad (9.11)$$

where r_{11} is the reliability of the original test,
and K is the multiplication factor for lengthening (or shortening) the test.

As an example, assume you have constructed a test of 20 items and have obtained a reliability estimate of 0.60. You are interested in estimating the reliability of the test if you were to double the number of items with items that are similar in inter-correlations, means and variances with the original 20 items. The factor K is 2 since you are doubling the length of the test. Your estimate would be:

$$R_{kk} = \frac{(2)(0.60)}{1 + (2 - 1)(.60)} = 0.75$$

Therefore, doubling the length of your test would result in an estimated reliability of 0.75, a sizable increase above the original 0.60. The formula can also be used to estimate the reliability of a shortened test constructed by sampling items from a longer test. For example a test of 100 items with a reliability of 0.90 could be used to produce a 25 item short-form test. The reliability would be

$$R_{kk} = \frac{(0.25)(0.90)}{1 + (0.25 - 1)(0.90)} = 0.6923$$

Note that in this case K = 0.25 since the test length has been changed by a factor of one fourth of the original length.

The Spearman-Brown formula can also be used to estimate the effects on a validity coefficient when either the test or the criterion measure have been extended in length. First we note that if a test is extended in length indefinitely (infinite length) then the reliability approaches 1.0. This permits us to estimate the validity between two measures, either or which (or both) have been extended in length. For example, the correlation between a test that has been extended by a factor of K and another test that has been extended by a factor of L is given by:

$$R_{KL} = \frac{r_{1I}}{\sqrt{1/K + (1 - 1/K)r}\sqrt{1/L + (1 - 1/L)r_{II}}} \quad (9.12)$$

where r_{1I} is the correlation between the two tests, r_{11} and r_{II} are the reliabilities of the two tests and K and L are the factors for extending the two tests.

If only one of the tests, say for example test I above, is made infinitely long so that its reliability approaches 1.0, then the above formula reduces to

$$R_{1\infty} = \frac{r_{1I}}{\sqrt{r_{II}}}$$

The above formula is useful in estimating the validity of a test correlated with a criterion measured without error. In addition, we may be interested in estimating the correlation of a test and criterion both of which have been adjusted for unreliability. This would estimate the correlation between the True scores of each instrument and is given by

$$R_{\infty\infty} = \frac{r_{1I}}{\sqrt{r_{11}r_{II}}} \quad (9.13)$$

Composite Test Reliability

Teachers often base course grades on the basis of a combination of tests administered over the period of the semester. The teacher usually, however, desires to give different weights to the tests. For example, the teacher may wish to weight tests 1, 2 as 1/4 of the total grade and the final exam (test 3) as 1/2 of the grade. Since the tests may vary considerably in length, mean, variance and reliability, one cannot simply add the weighted raw scores achieved by each student to get a total score. Doing so would give greater weight than intended to the more variable test and less weight than intended to the less variable test. A preferable method of obtaining the total weighted score would be first to standardize each test to a common mean and standard deviation. This is usually done with the z score transformation, i.e.

$$z_i = \frac{(X_i - \bar{X})}{S_x} \quad (9.14)$$

Each subject's z score for a test may then be weighted with the desired test weight and the sum of the weighted z scores be used as the total score on which grades are based. The reliability of this composite weighted z score can be estimated by the following formula:

$$R_{ww} = \frac{WCW'}{WRW'} \qquad (9.15)$$

where R_{ww} is the reliability of the composite,
W is a row vector of weights and W' is the column transpose of W,
R is the correlation matrix among the tests and
C is the R matrix with the diagonal elements replaced with estimates of the individual test reliabilities.

As an example, assume a teacher has administered three tests during a semester course and obtains the following information:

Test	Correlations		
	1	2	3
1	1.0	0.6	0.4
2	0.6	1.0	0.5
3	0.4	0.5	1.0
Reliability	0.7	0.6	0.8
Weights	0.25	0.25	0.50

The reliability of the composite score would then be obtained as:

$$R_{ww} = \frac{(.25\ .25\ .50)\begin{vmatrix}.7 & .6 & .4\\ .6 & .6 & .5\\ .4 & .5 & .8\end{vmatrix}\begin{pmatrix}.25\\ .25\\ .50\end{pmatrix}}{(.25\ .25\ .50)\begin{vmatrix}1.0 & .6 & .4\\ .6 & 1.0 & .5\\ .4 & .5 & 1.0\end{vmatrix}\begin{pmatrix}.25\\ .25\\ .50\end{pmatrix}}$$

$$= 0.861$$

The above equation utilizes matrix multiplication to obtain the solution. If you have not used matrix algebra before, you may need to consult an elementary text book in matrix algebra to familiarize yourself with the basic operations.

Reliability by ANOVA

Sources of Error: An Example

In the previous sections, an observed score for an individual on a test was considered to consist of two parts, true score and error score, i.e. $X = T + E$. Error scores were assumed to be random with a mean of zero and uncorrelated with the true score.

Reliability by ANOVA

We now wish to expand our understanding of sources of errors and introduce a method for estimating components of error, that is, analyzing total observed score variance into true score variance and one or more sources of error variance. To do this, we will consider a measurement example common in education—the rating of teacher performance.

A Hypothetical Situation

Assume that teachers in a certain school district are to be rated by one or more supervisor one or more times per year. Also assume that a rater employs one or more "items" in making a rating, for example, lesson plan rating, handling of discipline, peer relationships, parent conferences, grading practices, skill in presenting material, sensitivity to students, etc.. We will assume that the teachers are rated on each item using a scale of 1–10 points with 1 representing very inadequate to 10 representing very superior performance. We note that in this situation:

1. Teachers to be rated are a sample from a population of teachers,
2. Supervisors doing the rating are a sample of supervisors,
3. Items selected are a sample of possible teacher performance items,
4. Ratings performed are a sample of possible replications, and
5. Teacher performance on a specific item may vary from situation to situation due to variation in teacher mood, alertness, learning, etc. as well as due to situational variables such as class size, instructional materials, time of day, etc..

We are interested of course in obtaining ratings which accurately reflect the true competence of a teacher and the true score variability among teachers (perhaps to reward the most meritorious teacher, identify teachers needing assistance, and selection of teachers for promotion). We must recognize however, a number of possible sources of variance in our ratings—sources other than the "true" competence of the teachers and therefore error of measurement:

(a) Variability in ratings due to items sampled from the population of possible items,
(b) Variability in ratings due to the sample of supervisors used to do the ratings,
(c) Variability in ratings due to the sample of teachers rated,
(d) Interactions among items, teachers and supervisors.

Let us assume in our example that six teachers are rated by two supervisors (principal and coordinator) on each of four items. Assume the following data have been collected:

	Principal	Coordinator	Combined
Item	1 2 3 4	1 2 3 4	Princ. Coord. Both

Teacher

1	9 6 6 2	8 2 8 1	23 19 42
2	9 5 4 0	7 5 9 5	18 26 44
3	8 9 5 8	10 6 9 10	30 35 65
4	7 6 5 4	9 8 9 4	22 30 52
5	7 3 2 3	7 4 5 1	15 17 32
6	10 8 7 7	7 7 10 9	32 33 65
SUM	50 37 29 24	48 32 50 30	140 160 300

Item Sums for Principal + Coordinator
98 69 79 54

We now define the following terms to use in a three way analysis of variance:
X_{ijk} = the rating for teacher i on item j from supervisor k.

$$\sum_{i=1}^{6}\sum_{j=1}^{4}\sum_{k=1}^{2}(X_{ijk})^2 = 2,214 \quad \text{Sum of Squares of single}$$

$$\sum_{j=1}^{4}\sum_{k=1}^{2}(X_{\cdot jk})^2 = 12,014 \quad \text{Sum of Squares over teachers.}$$

$$\sum_{i=1}^{6}\sum_{k=1}^{2}(X_{i\cdot k})^2 = 8,026 \quad \text{Sum of Squares over items.}$$

$$\sum_{i=1}^{6}\sum_{j=1}^{4}(X_{ij\cdot})^2 = 4,258 \quad \text{Sum of Squares over supervisors.}$$

$$\sum_{k=1}^{2}(X_{\cdot\cdot k})^2 = 45,200 \quad \text{Sum of Squares over teachers and items.}$$

$$\sum_{j=1}^{4}(X_{\cdot j\cdot})^2 = 23,522 \quad \text{Sum of Squares over teachers and supervisors}$$

$$\sum_{i=1}^{6}(X_{i\cdot\cdot})^2 = 15,878 \quad \text{Sum of Squares over items and supervisors.}$$

$$(X_{\cdot\cdot\cdot})^2 = 90,000 \quad \text{Square of grand sum of all observations.}$$

Reliability by ANOVA

Our analysis of variance table may contain the following sums of squares:

$$\text{Total Sums of Squares} = \sum_{i=1}^{6}\sum_{j=1}^{4}\sum_{k=1}^{2}(X_{ijk})^2 - \frac{\left(\sum_{i=1}^{6}\sum_{j=1}^{4}\sum_{k=1}^{2}X_{ijk}\right)^2}{(6)(4)(2)}$$

or $SS_{total} = 2{,}214 - 90{,}000/48 = 339.00$

$$\text{Teacher Sums of Squares} = \frac{\sum_{i=1}^{6}(X_{i..})^2}{(4)(2)} - \frac{\left(\sum_{i=1}^{6}\sum_{j=1}^{4}\sum_{k=1}^{2}X_{ijk}\right)^2}{(6)(4)(2)}$$

or $SS_{teachers} = 15{,}878\ /\ 8 - 90{,}000\ /\ 48 = 109.80$

$$\text{Item Sums of Squares} = \frac{\sum_{j=1}^{4}(X_{.j.})^2}{(6)(2)} - \frac{\left(\sum_{i=1}^{6}\sum_{j=1}^{4}\sum_{k=1}^{2}X_{ijk}\right)^2}{(6)(4)(2)}$$

or $SS_{items} = 23{,}522\ /12 - 90{,}000/48 = 85.2$

$$\text{Supervisor Sum of Squares} = \frac{\sum_{k=2}^{2}(X_{..k})^2}{(6)(4)} - \frac{\left(\sum_{i=1}^{6}\sum_{j=1}^{4}\sum_{k=1}^{2}X_{ijk}\right)^2}{(6)(4)(2)}$$

or $SS_{superv} = 45{,}200\ /\ 24 - 90{,}000\ /\ 48 = 8.3$

$$\text{Teacher} - \text{Item Interaction} = \frac{\sum_{i=1}^{6}\sum_{j=1}^{4}(X_{ij.})^2}{2} - \frac{\left(\sum_{i=1}^{6}\sum_{j=1}^{4}\sum_{k=1}^{2}X_{ijk}\right)^2}{(6)(4)(2)}$$
$$- SS_{teachers} - SS_{items}$$

or $SS_{TxI} = 4{,}258/2 - 90{,}000/48 - 109.8 - 85.2 = 59.00$

$$\text{Teacher} - \text{Superv. Inter} = \frac{\sum_{i=1}^{6}\sum_{k=1}^{2}(X_{i.k})^2}{4} - \frac{\left(\sum_{i=1}^{6}\sum_{j=1}^{4}\sum_{k=1}^{2}X_{ijk}\right)^2}{(6)(4)(2)}$$
$$- SS_{teachers} - SS_{superv}$$

or $SS_{TxS} = 8,026/4 - 90,000/48 - 109.8 - 8.3 = 13.4$

$$\text{Item} - \text{Superv. Interact.} = \frac{\sum_{j=1}^{4}\sum_{k=1}^{2}(X_{.jk})^2}{(6)} - \frac{\left(\sum_{i=1}^{6}\sum_{j=1}^{4}\sum_{k=1}^{2}X_{ijk}\right)^2}{(6)(4)(2)} - SS_{items} - SS_{superv}$$

or $SS_{IxS} = 12,014/6 - 90,000/48 - 85.2 - 8.3 = 33.8$

$$\text{Teacher} - \text{Item} - \text{Super.} = \sum_{i=1}^{6}\sum_{j=1}^{4}\sum_{k=1}^{2}(X_{ijk})^2 - \frac{\left(\sum_{i=1}^{6}\sum_{j=1}^{4}\sum_{k=1}^{2}X_{ijk}\right)^2}{(6)(4)(2)} - SS_{teachers} - SS_{items} - SS_{superv}.$$

or $SS_{TxIxS} = SS_{total} - (SS_{teachers} + SS_{items} + SS_{superv} + SS_{TxI} + SS_{TxS} + SS_{IxS})$
$= 339.0 - (109.8 + 85.2 + 59.0 + 13.4 + 33.8) = 29.5$

The Analysis of Variance table may be summarized as:

SOURCE	D.F.	SS	MS
Teachers (T)	5	109.8	21.96
Items (I)	3	85.2	28.40
Supervisors (S)	1	8.3	8.30
T x I Interaction	15	59.0	3.93
T x S Interaction	5	13.4	2.68
I x S Interaction	3	33.8	11.27
T x I x S Inter.	15	29.5	1.97

We may now use each of the above mean squares to estimate population variance components in examining the reliability of the ratings. We have:

$$S^2_{TxIxS} = MS_{TxIxS} = 1.97$$

The second order interaction is our error (residual) term since we only have a single observation under each of the three facets (teachers, items and supervisors).

$$S^2_{TxI} = .5(MS_{TxI} - MS_{TxIxS}) = .5(3.93) - 1.97 = 0.98$$

This is our error variance attached to teacher interaction with items. Each mean square at a given level includes variance at a higher level of interaction. We subtract out that previously obtained portion. We also divide by the number of observations on which the term is based—in this case the teacher by item interaction is based on two supervisors.

$$S^2_{TxS} = (1/4)(MS_{TxS} - MS_{TxIxS}) = .25(2.68 - 1.97) = .18$$

This is our estimate of error due to interaction of teachers and supervisors (repeated over the four items).

$$S^2_{IxS} = (1/6)(MS_{IxS} - MS_{TxIxS}) = (11.27 - 1.97)/6 = 1.55$$

This is the estimated error variance for interaction of items and supervisors over the six teachers.

$$S^2_T = [1/(4)(2)][MS_T - MS_{TxI} - MS_{TxS} + MS_{TxIxS}]$$
$$= (21.96 - 3.93 - 2.68 + 1.97) / 8 = 2.16$$

This is our estimate of variance due to differences among teachers—that variance we hope is large in comparison to error variance. It is our estimate of the teachers variance component of each rating by each supervisor.

$$S^2_I = [1/(6)(2)][MS_I - MS_{TxI} - MS_{IxS} + MS_{TxIxS}]$$
$$= (28.4 - 3.93 - 11.27 + 1.97) / 12 = 1.26$$

This is variance due to variability of ratings among the items or item "difficulty."

$$S^2_S = [1/(6)(4)][MS_S - MS_{TxS} - MS_{IxS} + MS_{TxIxS}]$$
$$= (8.3 - 2.68 - 11.27 + 1.97) / 24 < 0$$

This estimate of variability due to supervisors is less than zero hence considered negligible. While variance cannot be less than zero, our small sample of supervisors that apparently rated quite consistently led to this estimate. Estimates may, of course, fall above or below the population values.

We now turn to the question of estimating the reliability of our ratings. In previous sections the classical definition of reliability was given as

$$r_{xx} = \frac{\sigma^2_{true}}{\sigma^2_{true} + \sigma^2_{error}} = \frac{\sigma^2_{true}}{\sigma^2_{observed}}$$

The "true" score variance for J items rated by K supervisors is given by

$$S^2_{true} = (JK)^2 S^2_T = [(4)(2)]^2 \; 2.16 = (64)(2.16) = 138.24$$

Our "observed score" variance is estimated by

$$S^2_{obs} = (JK)(JKS^2_t + JS^2_s + KS^2_i + JS^2_{TxS} + KS^2_{TxI} + S^2_{IxS} + S^2_{TxIxS})$$
$$= (4 \times 2)\,[(4 \times 2)2.16 + (4)0 + (2)1.26 + (4)0.18 + (2)0.98 + 1.55 + 1.97]$$
$$= 208$$

and the ratio $S^2_{true}/S^2_{observed} = r_{xx} = 0.665$ is the estimate of the correlation that would be obtained between two sets of scores for a group of teachers rated on the basis of a random set of four items chosen for each teacher and rated by a random set of two supervisors for that teacher. Note our emphasis that this is a random effects model—each teacher could be rated on a sample of different items and by different supervisors!

In examining the sources of error, increasing the number of items would most likely reduce the largest error components (items and interaction of items with teachers and supervisors).

If the items used by each person doing the ratings is the same (fixed effects of items), the variance component for items disappears from the estimate of observed score variance giving

$$S^2_{observed} = (0 + .24 + 0.09 + 0.19 + 0.25) = 2.93$$

and

$$r_{xx} = 2.16/2.93 = 0.74$$

Obviously, using the same test on all teachers yields a more precise estimate of the teacher competencies. If we also fix the supervisors so that all teachers are rated by the same two supervisors, then S^2_S and S^2_{IxS} disappears as sources of error variance and the observe score is given by

$$S^2 = S^2_T + S^2_{TxI}/J + S^2_{TxS}/K + S^2_{TxIxS}/JK$$
$$= 2.16 + 0.24 + 0.09 + 0.25 = 2.74$$

and

$$r_{xx} = 2.16\,/\,2.74 = 0.79$$

By using the same items and supervisors, the reliability of the ratings has been increased from 0.66 to 0.79.

We may further assume that our items are not a sample from a population of items but, in fact, constitute the universe of teacher behaviors to which we intend to generalize. In this case, S^2_{TxI} and S^2_I will both disappear from our error term. Our estimates of true and observe score therefore become:

$$S^2_{true} = S^2_T + S^2_{TxS}/J = 2.16 + 0.24 = 2.40$$

and

$$S^2_{observed} = S^2_{true} = (S^2_S/K + S^2_{TxS}/K + S^2_{IxS}/JK + S^2_{TxIxS}/JK) = 2.93$$

Therefore $r_{xx} = 2.4 / 2.93 = 0.82$

Finally, if we choose to consider only two specific supervisors as our universe of supervisors, then

$$S^2_{true} = S^2_T + S^2_{TxS}/J + S^2_{TxS}/K = 2.16 + .24 + .09 = 2.49$$

and

$$S^2_{observ} = S^2_{true} + S^2_{TxIxS}/JK = 2.49 + 0.25 = 2.74$$

Therefore, $r_{xx} = 2.49 / 2.74 = 0.91$

Clearly, the degree to which one intends to generalize a test or rating procedure affects the reliability of the measurements for that purpose.

In the previous discussion we have examined multiple facets of reliability. We saw that the assumptions of sampling both test items and raters as well as subjects affected our estimate of reliability. We now will relate the above analysis with a simple ANOVA approach using the "Treatments by Subjects" analysis of variance program found in the Measurement Menu of the OpenStat system. To illustrate its use, we will combine the two supervisor ratings from the above example and treat our data as consisting of six teachers who have been rated on four items. We assume we are using the population of "items" and the same raters on each teacher rated. Our data consists of the following:

Teacher	Item 1	2	3	4	Sum
1	17	8	14	3	42
2	16	10	13	5	44
3	18	15	14	18	65
4	16	14	14	8	52
5	14	7	7	4	32
6	17	15	17	16	65
Sum	98	69	79	54	300

In calculating the sums of squares for the ANOVA, we first obtain the squares of individual ratings, squares of the sums for each teacher, squares of the sums for each item and the square of the sum of the item (or teacher) sums. These are:

$$\sum_{i=1}^{6}\sum_{j=1}^{4}(X_{ij})^2 = 4,258 \quad \text{Squares of single observations}$$

$$\sum_{i=1}^{6}(X_{\cdot j})^2 = 15,878 \quad \text{Squares of teacher sums}$$

$$\sum_{j=1}^{4}(X_{\cdot j})^2 = 23,522 \quad \text{Squares of item sums}$$

$$(X_{\cdot\cdot})^2 = 90,000 \quad \text{Square of grand total}$$

The sum of squared deviations about the mean for the terms of our ANOVA are obtained using the above terms and computed as follows:

$$SS_{total} = 4,258 - 90,000/24 = 508$$

$$SS_{teachers} = 15,878/4 - 90,000/24 = 219.50$$

$$SS_{items} = 23,522/6 - 90,000/24 = 170.33$$

$$SS_{IxT} = SS_{total} - SS_{teachers} - SS_{items} + 90,000/24 = 118.17$$

The SS_{items} and SS_{IxT} are often combined into a SS_{within} to represent the total sum of squares due to variation within subjects, i.e. the squared deviations of subject's scores about the subject means. The ANOVA summary table may look as follows:

SOURCE	D.F.	SS	MS	F
Among Teachers	5	219.50	43.90	5.57
Within Teachers	18	288.50	16.03	
Items	3	170.44	56.78	7.21
Teachers x Items	15	118.17	7.88	
Total	23	508.00		

The terms for our reliability are

$$S^2_{true} = (MS_{observed} - MS_{TxI})/N$$
$$= (43.90 - 7.88)/6 = 6.00$$

$$S^2_{observed} = S_{true} + MS_{TxI}/N = 6.02 + 7.88/6$$
$$= 6.00 + 7.88/6 = 7.31$$

and the reliability is

$$r_{xx} = S^2_{true}/S^2_{observed} = 6.00/7.31 = 0.82$$

This reliability is called the adjusted average rating reliability on the printout from the program in your system. It reflects the reliability of ratings in which the error due to differences in average ratings by the judges or items has been removed. Essentially, the individual ratings are "adjusted" so that the column sums or means are equal. If a test of J dichotomously scored items are analyzed by both the Kuder-Richardson Formula 20 and the Treatment by Subjects ANOVA procedures, the KR#20 reliability will equal the reliability reported above.

One can also estimate a single item reliability by obtaining an average item reliability using

$$r_{single} = \frac{MS_T - MS_{TxI}}{MS_T + (J-1)MS_{TxI}} = \frac{43.9 - 7.88}{43.9 - (3)7.88} = 0.53$$

Again, this reliability reflects an adjustment for the "difficulty" of the items, that is, all ratings or items are made to reflect the same sum or average across the subjects rated. A similar result would be obtained by using the Spearman-Brown Prophecy formula where we estimate the reliability of a test reduced in length to a single item.

Should the user want to know what the reliability of the ratings or test is without adjustment for variability in mean ratings, then the following may be used:

For the unadjusted test reliability

$$r_{xx} = 1.0 - (MS_{within}/MS_T)$$
$$= 1.0 - (16.03 / 43.90) = 0.63$$

For the estimate of a single item reliability unadjusted for difference among item (or rating) means, the formula is

$$r_{xx} = \frac{MS_T - MS_{within}}{MS_T + (J-1)MS_{within}}$$
$$= (43.9 - 16.03)/(43.9 + (4-1)16.03)$$
$$= 0.30$$

Item and Test Analysis Procedures

Teachers typically construct their own tests to measure the achievement of students in their courses. In constructing the test, it is a good idea to begin with a test "blueprint" or table of specifications for the test. This test blueprint usually consists of a table in which the rows represent content or concept areas to be tested and the columns represent levels of thinking required such as classified by Bloom's taxonomy of cognitive skills. The cells may simply indicate the number of items to be written in each concept area at each level of thinking skill. For example, an elementary teacher might construct a blueprint for a test over arithmetic concepts for eighth grade students using something like the following:

Level Concept	Knowledge	Application	Synthesis	Evaluation
Addition	3	3	1	1
Subtraction	2	2	1	2
Multiplication	3	3	0	0
Division	2	2	2	1
Percentage	3	3	3	3
Exponents	3	3	1	1

In this example, the teacher would construct 47 items from the table of specifications. The items constructed may be of a variety of types such as multiple choice, matching, completion, problem solving, etc.. Once the test is constructed and administered to the students, the teacher may then evaluate various properties of the items and test. For example, the teacher may want to know how reliable the test is, how difficult each item was, how well each item differentiates between high and low scoring students, and how the test might be improved for subsequent use. This section describes several methods for analyzing tests and the items within tests.

Classical Item Analysis Methods

Item Discrimination

If a test is constructed to test one predominant domain or area of achievement or knowledge then each item of the test should correlate positively with a total score on the test. The total score on the test is usually obtained by awarding a value of 1 to a student if they get an item correct and a 0 if they miss it and summing across all items. On a 47 item test, a student that gets all items correct would therefore have a total score of 47 while the student that missed all items would have a score of 0.

We can correlate each item with the total score obtained by the students. We may use the Pearson Product–moment correlation formula (see the section on simple correlation and regression) to do our calculations. We note however that we are correlating a dichotomous variable (our item is scored 0 or 1) with a continuous variable (total scores vary from 0 to the number of items in the test). This type of correlation is also called a "Point-Biserial Correlation." Unfortunately, when one of the variables in the product–moment correlation is dichotomous, the correlation is affected by the proportion of scores in the dichotomous variable. If the proportion of 0 and 1 scores is about the same (50 % for each), the correlation may approach 1.0. When the split of the dichotomous variable is quite disproportionate, say 0.2

and 0.8, then the correlation is restricted to much lower values. This certainly makes interpretation of the point-biserial correlation difficult. Nevertheless, a "good" test item will have positive correlations with the total score of the test. If the correlation is negative, it implies that more knowledgable students are more likely to have missed the item and less knowledgeable students likely to have gotten the item correct! Clearly, such an item is inconsistent with the measurement of the remaining items. Remember that the total score contains, as part of the total, the score of each item. For that reason, the point-biserial correlation will tend to be positive. A "corrected" point-biseral correlation can be obtained by first subtracting the individual item score from the total score before calculating the correlation between the item and total. If a test has many items, say more than 30, the correction will make little difference in the correlation. When only a few items are administered however, the correction should be applied.

The point-biserial correlation between test item and test total score is a measure of how well the item discriminates between low and high achievement students. It is a measure of item discrimination potential. Other item discrimination indices may also be used. For example, one may simply use the difference between the proportion passing the item in students ranking in the top half on the total score and the proportion passing the item among students in the bottom half of the class. Another index, the biserial correlation, may be calculated if one assumes that the dichotomously scored item is actually an imprecise measure of a continuous variable. The biserial correlaiton may be obtained using the formula:

$$r_{bis} = r_{pbis} \sqrt{p_i q_i / y_I} \qquad (9.16)$$

where r_{pbis} is the point-biserial correlation, p_i and q_i are the proportions passing and failing the item, and y_i is the ordinate of the normal curve corresponding to the cumulative proportion p_i.

Item Difficulty

In classical test analysis, the difficulty of an item is indicated by the proportion of subjects passing the item. An easy item therefore has values closer to 1.0 while more difficult items have values closer to 0.0. Since the mean of an item scored either 0 or 1 is the same as the proportion of subjects receiving scores of 1, the mean is the difficulty of the item. An ideal yardstick has markings equally spaced across the ruler. This permits its use to measure objects varying widely in length. Similarly, a test composed of items equally spaced in difficulty permits reasonable precision in measuring subjects that vary widely in their knowledge. With item difficulties known, one can select items along the continuum from 0 to 1.0 so that the revised instrument has approximately equal interval measurement. Unfortunately, the sample of subjects on which the item difficulty estimates are based must adequately represent all of the subjects for which the instrument is to be used. If another group of subjects that differs considerably in their knowledge is used to estimate the item difficulties, quite different estimates can be obtained. In other words, the item difficulty estimates obtained in classical test analysis methods are dependent on the sample from which they are obtained. It would clearly be desirable to have item parameter estimates that are invariant from group to group, that is, independent of the subjects being measured by those items.

In our discussion we have not mentioned errors of measurement for individual items. In classical test analysis procedures we must assume that each item measures with the same precision and reliability as all other items. We usually assume that errors of measurement for single items are normally distributed with a mean of zero and that these errors contribute proportionally to the error of measurement of the total test score. Hence the standard error of measurement is assumed equal for subjects scoring from 1 to 50 on a 50 item test!

The Item Analysis Program

The OpenStat package includes a program for item analysis using the Classical test theory. The program provides for scoring test items that have been entered as 0s and 1s or as item choices coded as numbers or letters. If item choices are in your data file, you will be asked to enter the correct choice for each item so that the program may convert to 0 or 1 score values for each item. A set of items may consist of several independent sub-tests. If more than one sub-test exists, you will be asked to enter the sequence number of each item in the sub-tests. You may also elect to correct for guessing in obtaining total scores for each subject. Either rights-wrongs or rights—1/4 wrongs may be elected. Finally, you may weigh the items of a test to give more or less credit in the total score to various items. An option is provided for printing the item scores and sub-score totals. You may elect one of several methods to estimate the reliability of the scored items. The sub-test means and standard deviations are computed for the total scores and for each item. In addition, the point-biserial correlation of each item with each sub-score total is obtained. Item characteristic curves are optionally printed for each item. The curves are based on the sub-score in which the item is included. The proportion of subjects at each decile on the sub-score that pass the item is plotted. If a reasonably large number of subjects are analyzed, this will typically result in an approximate "ogive" curve for each item with positive point-biserial correlations. Examination of the plots will reveal the item difficulty and discrimination characteristics for various ability score groups.

Item Response Theory

The past few decades has seen a rapid advance in the theories of psychological measurement. Among the more important contributions is the conceptualization of subject's responses to a single item. Simply stated, we assume that the probability of a subject correctly answering an item is a function of both subject and item parameters (stable characteristics). Usually the subject is considered to have one parameter—ability (or knowledge). The item, on the other hand, may have one or more parameters. Item difficulty is one parameter but item discrimination and chance-correctness are two other possible parameters to estimate. For example, a multiple choice item with five alternatives has a smaller probability of being correctly answered by guessing than a true-false type of question. Additionally, some items may differentiate among a broad range of student abilities while others discriminate only among subjects within a narrow range of abilities.

The functional relationship between the probability for correctly answering a question and the ability of subjects is usually represented by an item characteristic curve such as that depicted below. We might use total scores on the test as approximations of subject's ability parameter and plot the proportion of subjects in each score group that correctly answered the item.

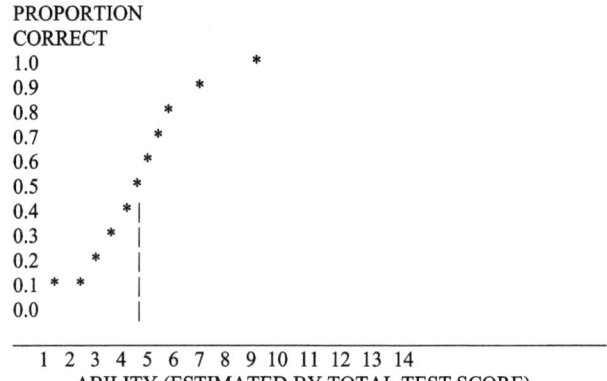

An individual's ability score may be obtained by averaging the probabilities for those items correctly answered and multiplying by the number of items in the test. In the figure above, a vertical line is drawn at the median (50 percentile). This represents the ability of subjects that have a 50–50 chance (odds) of passing the item. It also may be considered the difficulty of the item. Note that the probabilities of passing the item increase continuously as the total score (or ability) of the subjects increase. We say that the probability of passing the item is a monotonic increasing function of ability. Clearly, an item for which the probability of correctly answering the item decreased as subject abilities increased would not be a desirable item! The slope of the curve at the median denotes the "discriminating power" of the item. If the slope is steep, a small change in subject ability produces a relatively large change in the probability of correctly answering the item. A very shallow slope would imply a low ability of the item to differentiate among subjects widely varying in ability. Typically, an item with a steep slope will only have that steepness over a relatively small range of abilities. For that reason, one item is insufficient to measure abilities with precision over a wide range of abilities. One would ideally have an instrument composed of multiple items with steep (and equal) sloped characteristic curves that overlapped on the linear portions of the curves. The figure below might represent a four item test with items equally spaced in difficulty and equal in discrimination:

```
PROPORTION
CORRECT

1.00                   1   2   3   4
0.95                1   2   3   4
0.9              1   2   3   4
0.8             1   2   3   4
0.7            1   2   3   4
0.6           1   2   3   4
0.5          1   2   3   4
0.4         1   2   3   4
0.3        1   2   3   4
0.2       1   2   3   4
0.1      1   2   3   4
0.05  1   2   3   4
0.00 1   2   3   4
     ─────────────────────────────
      1 2 3 4 5 6 7 8 9 10 11 12 13
                  ABILITY
```

It is apparent that items 1, 2, 3 and 4 above provide a different amount of information concerning the ability of subjects that differ in their ability. For example, item one provides little information about subjects that have total score ability greater than 8. Similarly, item 4 provides little information for subjects scoring below 5. The amount of discrimination information of an item for varying levels of ability is a function of the slope of the item line at each ability level. If we can describe the rate of change of ability at any point on an item characteristic curve, we can plot that rate of change against ability level. Such "plots" are called item information curves. A test information curve can similarly be plotted by summing the item information (rate of ability change) at each ability level. For an item of moderate difficulty and relatively steep slope, such an item information function might look like the figure below:

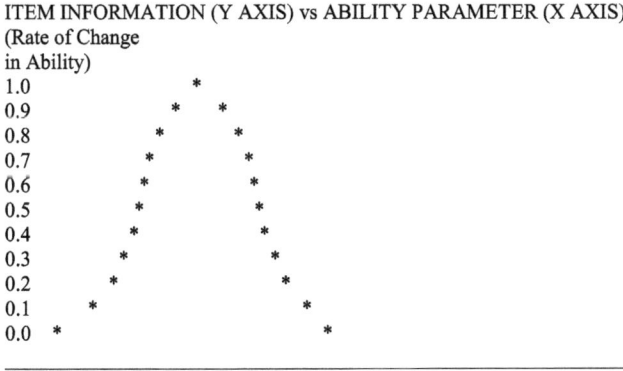

The One Parameter Logistic Model

In the classical approach to test theory, the item difficulty parameter is estimated by the proportion of subjects in some "norming" group that passes the item. Other methods may be used however, to estimate item difficulty parameters. George Rasch developed one such method. In his model, all items are assumed to have equal item characteristic slopes and little relevant chance probabilities. The probability of a subject answering an item correctly is given by the formula

$$P(X = 1|b_i) = \frac{e^{D(d_j - b_i)}}{1.0 - e^{D(d_j - b_i)}} \tag{9.17}$$

where b_i is the ability of an individual,
d_j is the difficulty of item j, D is an arbitrary scaling or expansion factor, and
e is the constant 2.7182818.....(the base of the natural logarithm system).

An individual's ability b_i is estimated by the product of the expansion factor D and the natural log odds of obtaining a score X out of K items, i.e.,

$$b_i = D \log \frac{X}{K - X} \tag{9.18}$$

The above equation may also be solved for X, the subject's raw score X expected given his or her ability, that is

$$X_i = \frac{K e^{(b_i / D)}}{1 + e^{(b_i / D)}} \tag{9.19}$$

The expansion factor D is a value which reflects the variability of both item difficulties d_j and abilities b_i. When scores are approximately normally distributed, this value is frequently about 1.7.

The Rasch one-parameter logistic model assumes that all items in the test that are analyzed measure a common latent variable. Researchers sometimes will complete a factor analysis of their test to ascertain this unidimensional property prior to estimating item difficulties using the Rasch model. Items may be selected from a larger group of items that "load" predominantly on the first factor of the set of items.

The OpenStat package includes a program to analyze subject responses to a set of items. The results include estimates of item difficulties in log units and their standard errors. Ability estimates in log units and errors of estimate are also obtained for subjects in each raw total score group. One cannot estimate abilities for subjects that miss all items or correctly answer all items. In addition, items that all subjects miss or get correct cannot be scaled. Such subjects or items are automatically eliminated by the program. The program will also produce item characteristic curves for each item and report the point-biserial correlation and the biserial correlation of each item with the total test score.

The Rasch method of calibrating item difficulty and subject ability has several desirable properties. One can demonstrate that the item difficulties estimated are independent of the abilities of subjects on which the estimates are based. For example, should you arbitrarily divide a large group of subjects into two groups, those who have total scores in the top half of the class and those who have scores in the bottom half of the class, then complete the Rasch analysis for each group, you will typically obtain item difficulties from each analysis that vary little from each other. This "sample-free" property of the item difficulties does not, of course, hold for item difficulties estimated by classical methods, i.e. proportion of a sample passing an item. Ability estimates of individuals are similarly "item-free." A subset of items selected from a pool of Rasch calibrated items may be used to obtain the same ability estimates of an individual that would be obtained utilizing the entire set of items (within errors of estimation). This aspect of ability estimation makes the Rasch scaled items ideal for "tailored" testing wherein a subject is sequentially given a small set of items which are optimized to give maximum precision about the estimated ability of the subject.

Estimating Parameters in the Rasch Model: Prox. Method

Item difficulties and subject abilities in the Rasch model are typically expressed in base e logarithm values. Typical values for either difficulties or abilities range between -3.0 and 3.0 somewhat analogous to the normally distributed z scores. We will work through a sample to demonstrate the calculations typically employed to estimate the item difficulties of a short test of 11 items administered to 127 individuals (See Applied Psychometrics by R.L. Thorndike, 1982, pages 98–100). In estimating the parameters, we will assume the test items involved the student in generating a response (not multiple choice or true false) so that the probability of getting the item correct by chance is zero. We will also assume that the items all have equal slopes, that is, that the change in probability of getting an item correct for a given change in student ability is equal for all items. By making these assumptions we need only solve for the difficulty of the item.

The first task in estimating our parameters is to construct a matrix of item failures for subjects in each total score group. A total score group is the group of subjects that have the same score on the test (where the total score is simply the total number of items correctly answered). Our matrix will have the total test score as columns and individual items as rows. Each element of the matrix will represent the number of students with the same total test score that failed a particular item. Our sample matrix is

The One Parameter Logistic Model

	Total test score										Total
Item	1	2	3	4	5	6	7	8	9	10	Failed
1	10	10	10	7	7	4	2	1	0	0	51
2	10	14	14	12	17	12	5	1	0	0	85
3	10	14	11	11	7	6	3	0	0	0	62
4	1	1	0	1	0	0	0	0	0	0	3
5	10	8	9	6	6	3	1	1	0	0	44
6	10	14	14	15	21	21	12	6	2	1	116
7	10	14	11	13	19	22	8	5	0	1	103
8	10	14	8	8	12	7	1	0	1	0	61
9	10	14	14	14	20	18	11	4	0	1	106
10	10	14	14	14	19	20	9	9	1	2	112
11	9	10	4	4	5	2	0	0	0	0	34
No.in Grp.	10	14	14	15	22	23	13	9	2	5	127

We begin our estimation of the difficulty of each item by calculating the odds of any subject failing an item. Since the far right column above is the total number of subjects out of 127 that failed the items in each row, the odds of failing an item are

$$\text{odds} = \frac{\text{no. failing}}{\text{no. subjects} - \text{no. failing}} \quad (9.20)$$

If we divided the numerator and denominator of the above ratio by the number of subjects we would obtain for any item i, the odds

$$\text{odds} = \frac{P_i}{1.0 - P_i} \quad (9.21)$$

Next, we obtain the natural logarithm of the odds of failure for each item. The mean and variance of these log odds are then obtained. Now we calculate the deviation of each item's log odds from the mean log odds of all items. To obtain the PROX. estimate of the item difficulty we multiply the deviation log odds by a constant Y. The constant Y is obtained by

$$Y^2 = \frac{1 + V/2.89}{1 - UV/8.35} \quad (9.22)$$

where V is the variance of the log odds of items and
W is the variance of the log odds of abilities.

Clearly, we must first obtain the variance of log odds for abilities before we can complete our PROX. estimates for items. To do this we must obtain the odds of subjects in each total score group obtaining their total score out of the total number of possible items. For subjects in each total score group the odds are

$$\text{odds} = \frac{\text{No. items passed}}{\text{No. items} - \text{No. items passed}} \quad (9.23)$$

For example, for subjects that have a total score of 1, the odds of getting such a score are $1/(11-1) = 1/10 = 0.1$.

Note that if we divide the above numerator and denominator by the number of test items, the formula for the odds may be expressed as

$$\text{odds} = \frac{P_j}{1 - P_j}$$

We obtain the logarithm of the score odds for each subject, and like we did for items, obtain the mean and variance of the log odds for all subjects. The variance of subject's log odds is denoted as U in the "expansion" factor Y above. A similar expansion factor will be used to obtain Prox. estimates of ability and is calculated using

$$X^2 = \frac{1 + U/2.89}{1 - UV/8.35} \tag{9.24}$$

The Prox. values for items is now obtained by multiplying the expansion factor Y (square root of the Y^2 value above) times the deviation log odds for each item. The Prox. values for abilities is obtained by multiplying the corresponding expansion value X times the log odds for each score group. The calculations are summarized below:

ITEM	FAILED	PASSED	ODDS	LOG ODDS	DEVIATION	PROX.
1	51	76	.67	-0.3989	-0.61	-0.87
2	85	42	2.02	0.7050	0.49	0.70
3	62	65	.95	-0.0473	-0.26	-0.37
4	3	124	.02	-3.7217	-3.93	-5.62
5	44	83	.53	-0.6347	-0.84	-1.20
6	116	11	10.55	2.3557	2.15	3.08
7	103	24	4.29	1.4567	1.25	1.79
8	61	66	0.92	-0.0788	-0.29	-0.41
9	106	21	5.05	1.6189	1.41	2.02
10	112	15	7.47	2.0104	1.80	2.58
11	34	93	.37	-1.0062	-1.22	-1.75

MEAN LOG ODDS DIFFICULTY = 0.21
VARIANCE LOG ODDS DIFFICULTY = 2.709

TOTAL SCORE	PASSED	FAILED	ODDS	LOG ODDS	PROX. ABILITY
1	1	10	.10	-2.30	-3.93
2	2	9	.22	-1.50	-2.56
3	3	8	.38	-0.98	-1.71
4	4	7	.	-0.56	-0.94
5	5	6	.83	-0.18	-0.31
6	6	5	1.20	0.18	0.31
7	7	4	1.75	0.56	0.94
8	8	3	2.67	0.98	1.71
9	9	2	4.50	1.50	2.56
10	10	1	10.00	2.30	3.93

MEAN LOG ODDS ABILITY = -0.28
VARIANCE LOG ODDS ABILITY = 1.038

Y EXPANSION FACTOR = 1.4315
X EXPANSION FACTOR = 1.709

Theoretically, the number of subjects in total score group j that pass item i are estimates of the item difficulty di and the ability bj of subjects as given by

$$b_j - d_i = \log[p_{ij}/(n_j - p_{ij})]$$

where p_{ij} is the proportion of subjects in score group j that pass item i and n_j is the number of subjects in score group j. The Prox. estimates of difficulty and ability may be improved to yield closer estimates to the p_{ij} values through use of the Newton-Rhapson iterations of the maximum-likelihood fit to those observed values. This solution is based on the theory that

$$p_{ij} = \frac{e^{(b_j - d_i)}}{1 + e^{(b_j - d_i)}} \tag{9.25}$$

It is possible, using this procedure, that values do not converge to a solution. The Rasch program included in the statistics package will complete a maximum of 25 iterations before stopping if the solution does not converge by that time.

If the Rasch model fits the data observed for a given item, the success and failure of each score group on an item should be adequately reproduced using the estimated parameters of the model. A chi-squared statistic may be obtained for each item by summing, across the score groups, the sum of two products: the odds of success times the number of failures and the odds of failure times the number of successes. This chi-squared value has degrees of freedom N—n where N is the total number of subjects and k is the total number of score groups. It should be noted that subjects with scores of 0 or all items correct are eliminated from the analysis since log odds cannot be obtained for these score groups. In addition, items which are failed or passed by all subjects cannot be scaled and are eliminated from the analysis.

Item Banking and Individualized Testing

Item banks are repositories of test questions in machine readable form. Typically, objective types of items and their choices are stored. For example, multiple choice, true-false, matching, incomplete sentences, and other types of items are stored. Each item consists of a "stem" and "foils". The stem is the major part of the question and the foils are the alternatives from which the examinee is to choose. The item bank must contain the "key", that is, the correct choice or weights for each foil which reflects the degree of correctness. An item bank typically contains hundreds of items in a general area, for example, statistics but these items may be subdivided into smaller domains such as parametric, nonparametric, univariate, multivariate, etc. Each item in the bank therefore has a classification code field. The classification code is useful in retrieving items of a given sub-domain when generating a test. An item bank also typically contains for each item, one or more estimates of parameters for the item obtained from prior administration of the item. For example, the item mean, variance, classical difficulty (proportion passing the item), logistic difficulty (perhaps as obtained from the Rash program), discrimination index and guessing factor (proportion expected right by guessing.)

Once an item bank is created, it may be used for several purposes. One common application is known as "tailored testing." This refers to the administration of items of known difficulty to a single subject. An item of medium difficulty is usually administered first. If the examinee misses that item, another item, half as difficult, is administered. On the other hand, if the subject passes the first item another item more difficult is administered. By selecting the next item to administer on the basis of the response to each previous item, the program quickly "converges" to that set of items for which the examinee has approximately a 50–50 chance of being right or wrong. This permits a much faster estimate of the subjects ability since only a small portion of test items must be administered.

Item banks may also be used to generate "parallel" tests, that is, tests that are similar in difficulty level and content sampling. These tests may be individually administered directly to the examinee on the computer or the test may be printed and reproduced for group administration. Experiments which involve pre and post testing of knowledge often utilize parallel tests so that changes measured may be attributed to treatment effects, not differences in test difficulty or content coverage.

A teacher or test administrator must have the capability of recording a variety of test specifications for generating different tests from the same item bank. An item bank system therefore contains procedures by which a teacher specifies the number of items to be in a test, the type of items to include, the range of acceptable item difficulties, the mode of test presentation, and the media for presenting the test.

A program that administers items "live" to subjects on the computer must possess a number of characteristics. Individual item responses must be collected as well as the total score for the individual. These must be filed in such a manner that both items and subject scores may be analyzed and summarized. The program should provide the option of giving "feedback" during administration, for example, telling the examinee whether they got the item right and if not, what the correct choice was. Some tests must be strictly timed. The program which administers the test on the computer should therefore provide the option of displaying the item for a specified period of time and if not answered within that time, go on to the next question.

Measuring Attitudes, Values, Beliefs

The evaluator of training workshops is often as interested in how participants "feel" about their training as well as how much they have learned and retained. The testing theory presented above dealt primarily with the measure of knowledge and gave the methods for defining and testing the reliability and validity of those measures. In a similar manner, we may be interested in developing and administering instruments to measure such things as:

(a) Attitudes toward management
(b) Attitudes toward training experiences
(c) Attitudes toward protected classes (women, minorities)
(d) Aattitudes toward alternative work arrangements
(e) Attitudes toward safety codes and/or practices
(f) Attitudes toward personnel in other departments

It is generally recognized that the way people feel about each other, their work environment and their work characteristics are important to their productivity and longevity on the job. This section is devoted to helping the evaluator construct instruments to measure such attitudes.

Methods for Measuring Attitudes

Most of you have completed at least one questionnaire of the following type:

```
--------------------------------------------------------------------------------
                              THESIS RESEARCH
                                   SURVEY
```

DIRECTIONS:

      Listed below are ten statements about thesis research. Please indicate whether you agree or disagree with each statement. Circle the A if you tend to agree with the statement or circle the D if you tend to disagree with the statement. Do not spend too much time thinking about each statement. Use your first impression. GO AHEAD!

A D   1.   The research one does for his or her thesis may determine the line of research they pursue the rest of their life.

A D   2.   The only reason theses are required is because the current faculty had to do one in order to graduate.

A D   3.   Most theses make little contribution to the body of knowledge in a discipline.

A D   4.   A thesis can demonstrate your ability to be creative and thorough in conducting a research project.

A D   5.   Unless you almost have a major in statistics, its very difficult to complete a useful thesis.

A D   6.   Reading a thesis is right up there with reading a telephone book for pleasure.

A D   7.   Certain fields like clinical psychology, business and technology where the graduate is not going to be a college professor should not require a thesis.

A D   8.   Ten years after completing their degree, most students are ashamed of their thesis.

A D   9.   The whole master's program is aimed at preparing the student to use research; the thesis is simply evidence of having achieved that goal.

A D 10.   Many theses have had profound effect on subsequent research and
----

The question asked of you is this: "How do you *score* the responses given by an individual to this type of instrument?" Do you simply add the "agrees" to get a total score? What if some of the statements the subject agrees with are negative statements? Do you "reverse" the scoring for those items? How do you know which items are negative? Would a group of judges have the same opinion as yours as to which are positive or negative items?

Clearly, when measuring an attitude, there is no actual "correct" or "incorrect" response! In order to "score" an attitude instrument as that shown above, we must first establish the degree to which each item expresses an attitude that is favorable or unfavorable toward the "object" or topic for which the items are written. Some items when agreed with may give evidence of a very strong attitude toward the positive or the negative end of a continuum. If we can establish a *scale value* for each item that indicates the degree of "positiveness" toward the object, we can then use those scale values to score the responses of a subject. One of the ways of doing this is to use a group of "judges" to establish those scale values. The following illustrates an instrument used to garner the opinion of judges about the "positiveness" of the items in the previous instrument:

THESIS RESEARCH ATTITUDE INSTRUMENT
JUDGE EVALUATION FORM

DIRECTIONS:

You are being asked to determine the positiveness or negativeness each of the following items. To do this, you will rate each item on a scale ranging from 1 to 7 where 1 indicates highly negative to 7 which indicates highly positive. In order to have a common "frame of reference" for each item, assume that a graduate student has agreed with the statement, then rate how positive or negative that student is toward dissertation research. As an example, use the following item:

A D   Most theses in Education are irrelevant surveys of little importance.

Assuming the student has marked AGREE (the underlined A) with the statement, how positive or negative do you think he (or she) is? Make a mark on the scale below to indicate your answer.

Highly Negative			Neither Positive Or Negative			Highly Positive
1	2	3	4	5	6	7

PLEASE BEGIN!

1. The research one does for his or her thesis may determine the line of research they pursue the rest of their life.
   1 | 2 | 3 | 4 | 5 | 6 | 7

2. The only reason theses are required is because the current faculty had to do one in order to graduate.
   1 | 2 | 3 | 4 | 5 | 6 | 7

3. Most theses make little contribution to the body of knowledge in a discipline.
   1 | 2 | 3 | 4 | 5 | 6 | 7

4. A thesis can demonstrate your ability to be creative and thorough in conducting a research project.
   1 | 2 | 3 | 4 | 5 | 6 | 7

5. Unless you almost have a major in statistics, its very difficult to complete a useful thesis.

   1 | 2 | 3 | 4 | 5 | 6 | 7

6. Reading a thesis is right up there with reading a telephone book for pleasure.

   1 | 2 | 3 | 4 | 5 | 6 | 7

7. Certain fields like clinical psychology, business and technology where the graduate is not going to be a college professor should not require a thesis.

   1 | 2 | 3 | 4 | 5 | 6 | 7

8. Ten years after completing their doctorate, most students are ashamed of their thesis.

   1 | 2 | 3 | 4 | 5 | 6 | 7

9. The whole doctorate program is aimed at preparing the student for research; the thesis is simply evidence of having achieved that goal.

   1 | 2 | 3 | 4 | 5 | 6 | 7

10. Many theses have had profound effect on subsequent research and products.

    1 | 2 | 3 | 4 | 5 | 6 | 7

By analyzing the responses of a group of judges, the median or mean rating of those judges can be used to determine a scoring weight for each item that can be used in scoring the subjects for whom we wish to obtain an estimate of their attitude. One of the methods often used to analyze these judge's ratings is called the method of successive intervals (see Edwards, 1951). A computer program on you statistics disk permits you to analyze such responses. Consult the program manual for directions on its use.

Affective Measurement Theory

Most classroom teachers first learn to develop tests of achievement over the content which they are engaged to teach. These tests fall in what is known as the Cognitive Domain of testing. Two additional areas of testing are, however, often just as important. These areas are the Psychomotor Domain and the Affective Domain. The Psychomotor Domain includes testing of fine and gross motor coordination, strength and accuracy. The affective domain includes the measurement of attitudes, values and opinions of subjects. Typically, we are interested in measuring an attitude on one major "latent" variable such as an attitude toward school, an attitude toward minorities, an attitude toward some political issue, etc. In such cases, all of the items of the instrument used to measure this attitude are related, in some manner, to the major latent variable. In the following discussion, we will make this assumption of unidimensionality, that is, that all items are directly related to the same, underlying construct.

Thurstone Paired Comparison Scaling

A variety of item types have been developed to measure attitudes and values. Two major forms are used most commonly: (a) the agree/disagree format and (b) the "Likert" scale type involving a degree of agreement or disagreement, usually on a five or more point scale. In the case of agree/disagree statements, the subject is simply asked to indicate whether they agree or disagree to each statement listed. The statements are written to represent both positive or negative attitudes toward the object of the measurement. For example, if we were measuring an attitude toward "going to college" we might have the following statements:

1. College degrees are extremely important if your goal is to be a professional.
2. College graduates are snobish and have lost touch with humanity.
3. If you really want to make money, you can easily do so without going to college.
4. So many people are going to college, a college degree doesn't mean much any more.

If, on the other hand, we were using the Likert form of the statements, we will tell each subject to mark how strongly they agree (or disagree) with each statement using a scale such as

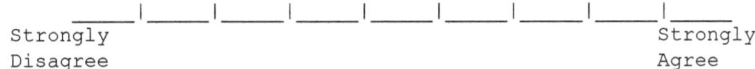
```
Strongly                                    Strongly
Disagree                                      Agree
```

You can see by the nature of the items, that there is no "correct" or "incorrect" response to each statement. Since we have no clear right or wrong answer, this poses a problem for "scoring" the responses of the instrument and obtaining a measure of the subject's attitude. We could arbitrarily mark those items which we feel reflect a positive attitude as a + 1 if the subject "agreed" with the statement (or marked closer to the agree on a Likert scale), and score 0 if they failed to agree to a positive item. For negatively stated items we could similarly score a subject as 0 if they agreed with the negative item and score them a + 1 if they disagreed with the negative item. The sum of these individual item scores, like our cognitive tests, would be the measure of the subject's attitude. Unfortunately, what you perceived as a "negative" or "positive" item may not be what I see for the same item! In fact, a group of judges might vary considerably in how "negative" or "positive" they felt each statement was toward the attitude object. Because of the ambiguity of attitude statements and because we desire to produce measurements for subjects which fit at least an interval scale of measurement, a variety of methods have been developed to "scale" the items used in affective instruments.

One of the first methods developed to determine the score values of items that subjects are asked to agree or disagree with is known as the Thurstone Paired-Comparisons Scaling method. This method utilizes a group of judges who are asked to compare each statement with every other statement and simply indicate which statement in each pair is more favorable toward the object if a subject were to agree with each one. For example, item 1 and item 2 of the above examples would be compared. If a judge felt that agreeing with item 1 indicated more favorableness toward going to college than agreeing with item 2, he would indicate item 1 is more favorable. By employing a reasonably large (say $N > 20$) number of judges, an average of the number of times judges selected each item over another can be obtained. If we assume these judgments by the judges are normally distributed around the "stimulus value" of each item, that is, the degree of favorableness of the items, we can obtain an estimate of the stimulus value for each item.

Let's consider an example of directions for the above 4 items that might be given to 30 judges:

```
DIRECTIONS:  Listed above are four statements which reflect
varying degrees of positiveness toward attending college.
Please indicate to the left of each pair of statements,
which item you feel reflects a more positive attitude toward
attending college.
_____   A.  Item 1     B.  Item 2
_____   A.  Item 1     B.  Item 3
_____   A.  Item 1     B.  Item 4
_____   A.  Item 2     B.  Item 3
_____   A.  Item 2     B.  Item 4
_____   A.  Item 3     B.  Item 4
```

Following administration of the above to 30 judges, we might obtain the following matrix. The number in the cells of this matrix reflect the number of judges which felt the item listed at the top was MORE favorable than the item listed to the left.

Judgement Matrix

ITEM	1	2	3	4
1	10	1	3	7
2	19	10	18	16
3	17	2	10	13
4	13	4	7	10

Notice in the above matrix that the diagonal values represent a comparison of a single item with itself. Since such comparisons are not actually made, we assume that one half of the time the item would be judged more positive and one half the time less positive. Also note that the values below the diagonal are the number of judges in the sample minus the value for the corresponding items above the diagonal.

To obtain the "scale value" of each item, we next convert the numbers of the above matrix first to the proportion of total judges and then we convert the proportions to z scores under the unit normal distribution. The matrices corresponding to the above example would be:

Proportion of Judgements

ITEM	1	2	3	4
1	.50	.05	.15	.35
2	.95	.50	.90	.80
3	.85	.10	.50	.15
4	.65	.20	.85	.50

z Scores for Proportions of Normal Curve

ITEM	1	2	3	4
1	0.00	-1.65	-1.04	-0.39
2	1.65	0.00	1.28	0.84
3	1.04	-1.28	0.00	-1.04
4	.39	-.84	1.04	0.00
Sum	3.08	-3.77	1.28	-.59
Average	.77	-.94	.32	-.15
Scale Value	1.71	0.00	1.26	.79

The last three rows above are simply the column sums, the column average, and the average plus the absolute value of the smallest column average. Since we are constructing a psychological scale, the mean and standard deviation of the scale values is arbitrary. We simply desire to build estimates of the intervals among the stimuli (items). The last row is labeled Scale Value. It reflects the average difference of the distance of each item from the other items on our psychological scale. The item (number 2) with the lowest scale value is the one which is "most negative" toward attending college. The item (number 1) with the largest value is the one most positive toward attending college. The scale values reflect the discriminations of the judges, NOT their attitudes. We simply used the judges to acquire "weights" for each item that reflect the degree of positivism or negativism of each item! Now that we have these scale values however, we can use them to actually measure the attitude of subjects toward attending college. To do this, our subjects would receive instructions something like

```
Directions:  Each statement below reflects an attitude
             about college.  You are to circle the A if
             you agree with the statement or circle the D
             if you disagree with the statement. Go ahead.


        A    D    1.   College degrees are extremely important
                       if your goal is to be a professional.
        A    D    2.   College graduates are snobish and have
                       lost touch with humanity.
        A    D    3.   etc.
```

Once a subject has indicated agreement or disagreement with the items, the subject's total score is calculated by simply averaging the scale value of those items with which they agreed. The Paired-Comparisons procedure described above makes several assumptions. First, it assumes that the judges discriminations among the items are normally distributed. Secondly, it assumes that the variance of those discriminations are equal. Third, it assumes that the items all measure, to varying degrees, the same underlying attitude. Fourth, it assumes that the correlation among the judges discriminations for item pairs are all equal. Fifth, it assumes the mean and standard deviation of the scale values are arbitrary and the scale reflects only distances among items, not absolute amounts of an attitude.

You have probably already noticed that if you have very many items, the number of item pairs that judges are required to judge becomes large. The number of unique pairs is obtained by $k(k-1)/2$ where k is the number of items. For example, if you constructed 20 statements, the judges would have to make $20(19)/2 = 190$ discriminations! Obviously you will try the patience of judges if your instrument is very long. A more convenient method of estimating item scale values is described in the next section.

Incidentally, if an item is judged to be higher than all other items by all judges or lower than all items by all judges, you would end up with a proportion of 1.0 or 0.0. The z scores corresponding to those proportions is plus or minus infinity and therefore could not be used to obtain an average. Such items may simply be eliminated or the obtained proportions changed to something like 0.99 or 0.01 as estimates of "what they might have been" if you had a much larger sample of judges.

Successive Interval Scaling Procedures

The Paired-Comparisons procedure described in the last section places great demands on judges if the number of items in an affective instrument is large. Yet we know that instruments with more items tend to give a more reliable estimate of an individual's attitude. The Successive Intervals scaling procedure provides a means of obtaining judges discriminations for k items in k judgments. The resulting scale values of items judged by both the Paired-Comparisons and Successive intervals methods correlate quite highly.

In the successive intervals scaling method, judges are asked to categorize statements on a continuum of an attribute like favorable-unfavorable. Typically five to nine categories are used, always using an odd number of categories. Utilizing the example from the previous section in which we are scaling items for measuring subjects attitudes toward attending college, a sample instruction to judges might look like the following:

Directions: Each item below reflects some degree of favorableness or unfavorableness toward attending college. Indicate the degree of favorableness in each item by making a check in one of the seven categories ranging from highly unfavorable to highly favorable.

1. College degrees are extremely important if your goal is to be a professional.

   |____|____|____|____|____|____|____|
   Highly   Highly
   Unfavorable   Favorable

2. College graduates are snobbish and have lost touch with humanity.

   |____|____|____|____|____|____|____|
   Highly   Highly
   Unfavorable   Favorable

3. etc.

If we assume again that we have a reasonably large sample of judges evaluating each item of our instrument, and we assume that the classifications of items on the continuous scale tend to be normally distributed, we employ computations similar to the Paired-Comparison method for estimating scale values. For our example above, we might obtain, for the group of judges, the following classifications:

Frequency of Item Classifications

Category:	1	2	3	4	5	6	7
Item							
1	0	1	1	3	8	6	1
2	2	7	6	4	1	0	0
3	1	3	6	6	3	1	0
4	1	5	9	4	1	0	0

To obtain scale values by the method of successive intervals, we next obtain the cumulative frequencies within each item, convert those to cumulative proportions, and then convert the cumulative proportions to z scores. For example:

Cumulative Frequencies and Proportions

Category Item:		1	2	3	4	5	6	7
1	cf	0	1	2	5	13	19	20
	cp	0	.05	.10	.25	.65	.95	1.0
2	cf	2	9	15	19	20	20	20
	cp	.10	.45	.75	.95	1.0	1.0	1.0
3	cf	1	4	10	16	19	20	20
	cp	.05	.20	.50	.80	.95	1.0	1.0
4	cf	1	6	15	19	20	20	20
	cp	.05	.30	.75	.95	1.0	1.0	1.0

Methods for Measuring Attitudes

z Score Equivalents to Cumulative Proportions

Category Item	1	2	3	4	5	6	7
1	–	-1.65	-1.28	-0.67	0.38	1.65	–
2	-1.28	-.13	.68	1.65	–	–	–
3	-1.65	-.85	.00	.85	1.65	–	–
4	-1.65	-.52	.68	1.65	–	–	–

Differences Between Adjacent Categories

Difference Item:	2-1	3-2	4-3	5-4	6-5	7-6
1		.37	.61	1.05	1.27	
2	1.15	.81	.97			
3	.80	.85	.85	.80		
4	1.13	1.20	.97			
Sum	3.08	3.23	3.40	1.85	1.27	
N	3	4	4	2	1	
Mean	1.03	.81	.85	.93	1.27	
Cum. Avg.	1.03	1.84	2.69	3.62	4.89	

Scale values for the items which have been judged and analyzed by the method of successive intervals is obtained using the formula for the median of an interval, that is:

$$\text{Scale Value} = \text{LL} + \frac{(.5 - \Sigma Pb)}{Pw}\bar{W} \tag{9.26}$$

where LL is the lower limit of the interval,
Pb is the Probability below the interval,
Pw is the Probability in the interval,

and $\bar{W}$ is the average interval width.

The scale value of items is the median value of the item on the scale defined by the cumulative average of the mean z score differences between categories. The scale values for the example above are therefore obtained as follows:

Scale value for item 1:

1. First, find the category in which the cumulative proportion is just less than 0.50, that is, that category just below the category in which the cumulative proportion is 0.5 or greater. For item 1 this is the category 4 (cumulative proportion = 0.25).
2. Next, obtain the cumulative average scale value for the category difference of the category just identified and the one below it. In this case, the cumulative average for the difference 4–3 which is 2.69. This represents the lower limit of the category in which the scale value for item 1 exists.

3. The value of EPb is the cumulative proportion up through the category identified in step (1) above, that is, 0.25 in our example.
4. The Pw is the proportion within the interval in which the median is found. In our example, it is the proportion obtained by subtracting the proportion up to category 5 from the proportion in category 5, that is, $0.65 - 0.25 = 0.40$.
5. Obtain the width of the interval next. This is the average z score differences in the interval in which the median is found. In this case the interval difference 5–4 has an average width of 0.93.
6. Substitute the values obtained in steps (1)–(5) in the equation to obtain the item scale value. For item 1 we have

$$S1 = 2.69 + \frac{(.50 - .25)}{(.65 - .25)} 0.93 = 3.2700$$

In a similar manner, the scale values for items 2 through 4 are:

$$S2 = 1.03 + \frac{(.50 - .45)}{(.75 - .45)} 0.81 = 1.1650$$

$$S3 = 1.03 + \frac{(.50 - .20)}{(.50 - .20)} 0.81 = 1.8343$$

$$S4 = 1.03 + \frac{(.50 - .30)}{(.75 - .30)} 0.81 = 1.3900$$

Several points should be made concerning the above computations. First note that the initial seven categories that were used represent midpoints of intervals. The number of judges placing an item within each category are assumed to be distributed uniformly accross the interval represented by the midpoint (category number). The calculation which involves subtracting the z scores in one category from those in the next higher category, and then averaging those values, establishes the distance between the midpoints of our original categories. In other words, there is no assumption of equal widths—we in fact estimate the interval widths. Once the interval widths are estimated, the accumulation of those widths describes the total scale of our measurements. You will have noticed that if the total number of categories is originally k (7 in our example), there will be k-2 differences obtained for adjacent categories. We have no way of estimating the width of the first and last category since there are no values below or above them. We can see this if we draw a schematic of the scale:

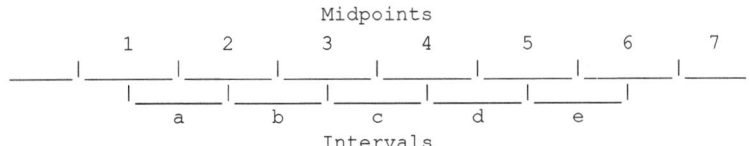

Methods for Measuring Attitudes

We can illustrate where each item lies on the obtained scale by "plotting" the scale value of each item:

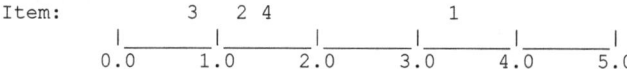

```
Item:            3    2 4                    1
        |_____|_____|_____|_____|_____|
       0.0      1.0     2.0      3.0      4.0      5.0
```

We can see that item 1 was judged more positive than the other three items and lies considerably further from the other items. Items 2, 3 and 4 are more similar in scale value with item 3 being judged the most negative of the four items.

Once the scale values of items are known, the same practice as employed in Paired-Comparisons methodology is used to obtain measures of individuals. The statements are presented to the subjects and the scale values of those items to which the subject agrees is averaged. The obtained average reflects the attitude of the subject.

Guttman Scalogram Analysis

If the items used to measure an attitude are all reflective of the same underlying attitude but to varying amounts, then subjects that vary on that attitude should agree or disagree to the items in a specific pattern. As an example, assume we have five items which measure the degree of positivism toward maintaining U.S. troops in a base in Japan. Now assume that these items are ranked in the order to which they evoke an "agree" response by six people that vary in their attitude toward maintaining the troops in Japan. If there is consistency of measurement, and we assign a "1" if a subject "agrees" and "0" if the subject "disagrees" with an item, we would expect that the following matrix of observations might be recorded:

	Rank of Item on the Attitude						
	1	2	3	4	5	Score	Rank
Subject							
1	1	1	1	1	1	5	1
2	0	1	1	1	1	4	2
3	0	0	1	1	1	3	3
4	0	0	0	1	1	2	4
5	0	0	0	0	1	1	5
6	0	0	0	0	0	0	6

In our example, subject 1 has agreed with all five statements and subject 6 has disagreed with all items. Note the items have been arranged in order from most negative toward maintaining troops to most positive toward retaining troops in Japan. In addition, the subjects have been arranged from the subject with the most positive attitude down to the subject with the least positive (most negative) attitude. The matrix of the responses reflects perfect agreement or order of the responses. In "real" life, we seldom get such a perfect pattern of responses. A more typical response pattern might look more like:

```
               Items Ordered by Total "Agree" Responses
                  1         2         3         4         5
    Response     1 0       1 0       1 0       1 0       1 0      Score

    Subject
       1         x         x         x         x         x          5
       2           x       x         x         x         x          4
       3         x           x       x         x         x          4
       4           x       x           x       x         x          3
       5           x         x         x       x         x          2
       6           x       x       x             x       x          2
       7           x       x         x           x       x          1
       8           x       x         x       x                      1
       9           x       x         x           x         x        0
      10           x       x         x           x         x        0

    sums         2 8       3 7       4 6       6 4       7 3
    Proportion  .2 .8     .3 .7     .4 .6     .6 .4     .7 .3
```

In this sample of ten subjects, we have several subjects with the same total score as another subject but a different pattern of "agree" or "disagree" to the statements. There is not perfect agreement among the items in differentiating the attitudes of the subjects! Note that we have recorded the response of each subject in one of two columns beneath each item. The sum or proportion of the "agree" or 1 responses is totaled accross subjects to identify the order of the "positivism" of the item. Item 5 is the item which received the greatest number of "agree" responses while item 1 received the fewest.

If we have "perfect" reproducibility in an instrument of k items, we would be able to perfectly reproduce the individual item responses of an individual given their total score (number of items to which they agree). If their is inconsistency of measurement, we can only estimate the likely response to each item. In order to make such estimates, it is necessary to identify a "cutting" point for each item which identifies that point where the pattern of agree/disagree responses most likely changes. This point is one where the number of errors is a minimum. An error is counted whenever a subject below the cutting score agrees with a statement or whenever a subject above the cutting point disagrees with the statement. For the above table, we have inserted the cutting scores which give the minimum error counts:

```
               Items Ordered by Total "Agree" Responses
                  1         2         3         4         5
    Response     1 0       1 0       1 0       1 0       1 0      Score

    Subject
       1         x         x         x         x         x          5
       2           x       x         x         x         x          4
       3         x_          x       x__       x         x          4
       4           x       x___        x       x         x          3
       5           x         x         x       x__       x          2
       6           x       x       x             x       x          2
       7           x       x         x           x       x__        1
       8           x       x         x       x             x        1
       9           x       x         x           x         x        0
      10           x       x         x           x         x        0

    sums         2 8       3 7       4 6       6 4       7 3
    Proportion  .2 .8     .3 .7     .4 .6     .6 .4     .7 .3
    Errors       0 1       0 1       1 0       1 0       0 0      Σe=4
```

There are actually several choices for cutting scores on each item which minimize the sum of the errors. L. Guttman (see Edwards, p. 182) has developed a coefficient which expresses the degree of reproducibility of a set of items. It is obtained as one minus the proportion of errors in the total number of responses. For the above data, we would obtain the coefficient of reproducibility as

$$\text{Rep} = 1.0 - 4/50 = 0.92$$

Because the cutting scores in the above matrix may be made at several points, the response pattern expected of a subject with a given total score might vary from solution to solution. In order to obtain a method of setting cutting scores that is always the same and thus yields a means of accurately predicting a response pattern, Edwards (Edwards, pgs. 184–188) developed another method for obtaining cutting scores. This method is illustrated for the same data in the figure below:

```
           Items Ordered by Total "Agree" Responses
              1        2        3        4        5
Response     1 0      1 0      1 0      1 0      1 0       Score

Subject
  1           x        x        x        x        x          5
  2                    x        x        x        x          4
           ------------------------------------------
  3           x                 x        x        x          4
           ------------------------------------------
  4           x        x                 x        x          3
           ------------------------------------------
  5           x        x        x        x        x          2
  6           x        x        x                 x          2
           ------------------------------------------
  7           x        x        x                 x          1
           ------------------------------------------
  8           x        x        x        x                   1
  9           x        x        x                 x          0
 10           x        x        x                 x          0

sums         2 8      3 7      4 6      6 4      7 3
Proportion   .2 .8    .3 .7    .4 .6    .6 .4    .7 .3
```

In the above display of our sample data, we have used the proportion of 1 responses (agree) to draw our cutting points. For example, in item 1, 20% of the subjects agreed with the item. The cutting score was then drawn below 20% of all the responses (both agree and disagree). This procedure was used for each item. Errors are then counted whenever a response disagrees with the pattern expected. For example, both subjects 1 and 2 are expected to have a pattern of responses 1 1 1 1 1 but subject 2 has 0 1 1 1 1 as a pattern. One response disagreed with the expected so the error count is 1 for subject 2. Subject three is expected to have a response pattern of 0 1 1 1 1 but in fact has a response pattern of 1 0 1 1 1 . Since there are two items that disagree with the expect pattern, the error count for subject 3 is 2. A similar procedure is followed for each subject. The expected pattern for each total score is shown below along with the number of errors counted for subjects with those total scores:

```
Total Score    Expected Pattern    Subject    No. of Errors

    5           1  1  1  1  1        1             0
    4           0  1  1  1  1        2             0
                                     3             2
    3           0  0  1  1  1        4             2
    2           0  0  0  1  1        5             0
                                     6             2
    1           0  0  0  0  1        7             0
                                     8             2
    0           0  0  0  0  0        9             0
                                    10             0

                                              Σe = 8

          Rep = 1.0 - ( 8 / 50) = 0.84
```

$$\text{Rep} = 1.0 - (8/50) = 0.84$$

This computation of the coefficient of reproducibility is a measure of the degree of accuracy with which statement responses can be reproduced on the basis of the total score alone! It is this latter method with is used in the program GUTTMAN found in the OpenStat program. The proportion of subjects agreeing or disagreeing with each item affects the degree of reproducibility. If very large or very small numbers of subjects agree to an item, the reproducibility is increased. The minimal coefficient of reproducibility may be obtained by the larger of the two values (a) proportion agreeing or (b) proportion disagreeing with a statement and dividing by the number of items. In our example these values are 0.8, 0.7, 0.6, 0.6 and 0.7. The minimal marginal reproducibility is therefore

$$\frac{.8 + .7 + .6 + .6 + .7}{6} = 0.68$$

The response pattern corresponding to this model response pattern is 0 0 0 1 1. If we were to predict each subjects responses with this pattern and count errors, the coefficient of reproducibility would be 0.68! The Guttman Coefficient of reproducibility may be thought of as an index somewhat comparable to the reliability coefficient. A value of one would indicate a set of items that are fully consistent in measuring differences among subjects.

In the methods of paired comparison and successive intervals, we utilized a group of judges to estimate scale values for items. These scale values were then used to obtain the scores for subjects administered the statements. With the Guttman scaling method, we do not use judges but simply the responses of the subjects themselves as a basis for determining their attitude scores. We simply assign 1 to the item with which they agree and 0 to those with which they disagree. If the instrument has a high coefficient of reproducibility, then the total of the subject response codes, i.e. their total score, should be directly interpretable as a measure of their attitude. The subject's total score may be divided by the number of items to obtain the proportion of items to which the subject agreed. It is assumed that all items reflect a varying degree of positivism to the attitude object (e.g. troops in Japan) and therefore the subject's total score based on those items also reflects the subject's attitude. The scale value of each item is the cutting score for that item. In the above example, we may place the items on the scale as follows:

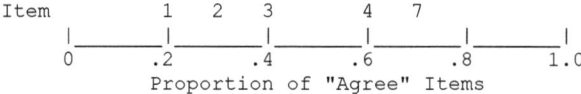

```
Item            1   2   3       4   7
|_____|_____|_____|_____|_____|
0       .2      .4      .6      .8      1.0
         Proportion of "Agree" Items
```

The items to which few subjects "agree" is a more negative item than the item to which a larger number of subjects agree. The proportion of items an individual subject agrees with is an indication of the subjects positivism toward the attitude object.

Likert Scaling

Also called the method of Summated Ratings, the Likert scaling method, like the Guttman method above, does not use judges to determine the scale value of items. Subjects are directly measured on each statement by indicating their degree of agreement, usually using a five-point scale. The statements administered are statements judged only by the person constructing the items as either a "favorable" or "unfavorable" item. If a five point scale is used such as

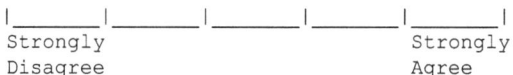

```
|_____|_____|_____|_____|_____|
Strongly                        Strongly
Disagree                        Agree
```

the lowest category is assigned a value of 0, the next category a 1, etc. up to the last category which would be assigned the value 4. If the item is an "unfavorable" item toward the attitude object, the category scores are reversed, that is, the first category assigned 4, the next 3, etc. To obtain a subject's score, one simply adds the values of the categories checked by the subject. Normal item analysis procedures may be used to eliminate items which do not measure the attitude consistent with other items. The point-biserial correlation of the item with the total score is the typical criterion used. If the item correlates quite low with the total score, the item should be eliminated.

It is important to note that the scores obtained by the Likert method cannot be interpreted without reference to a comparison group. Since the item scale values are not obtained, and the distances among the items is therefore unknown, the total scores are only meaningful in reference to a comparison group. For example, say that a scale of 20 items is administered to a subject and the subject's score is 5. This score cannot be directly interpreted. It may be that in one group of subjects this is a highly positive score while in another group, a very low score. We cannot say the score of 5, by itself, reflects a positive or negative attitude toward the object. It has been found in previous research that scores obtained on a Likert scale correlate quite high with the same items scaled and scored by the Thurstone method. If the interest of the researcher is to use the attitude measures to describe its relationship with some other variables through correlation methods, then the Likert method is cost-effective. If, on the other hand, the researcher desires to interpret individual attitudes as being positive or negative toward some object, then a method such as the paired-comparison or successive interval scaling method should be employed.

Semantic Differential Scales

Osgood, et al. (1971) developed a measure of the "meaning" attached, through a theorized learning model, to a variety of stimuli including both physical objects as well as "ideas" or concepts. Their measure is based, briefly, on the notion that certain words have become associated with subject's responses to objects through conditioning and generalization of conditioning. They observed that in many situatations, people, for example, might use words such as heavy, dark, gloomy to describe some classical music while words such as bright, up, shiny, happy might describe other music. These words which are also used to describe many objects appear to have general utility for subjects in describing their "feelings" about an object. Osgood and his colleagues utilized factor analysis procedures to identify subsets of items which appear to measure different dimensions of meaning. Their goal was to identify a set of bipolar adjectives which describes the "semantic space" of given objects. This space is described by orthogonal axis of the bipolar adjectives. The objects lie within this space at varying distance from the origin (intensity) and in specific directions (description). Three major dimensions of the semantic space are typically used. These are (I) Evaluation, (II) Activity, and (III) Potency.

The semantic differential scale is constructed of those bipolar adjectives (e.g. hot—cold) which are demonstrated to differentiate the meaning attached by individuals to a given object (e.g. school attendance). Thus the first problem in constructing a semantic differential scale is the selection of bipolar adjective pairs that measure predominantly one dimension of the semantic space and differentiate among individuals that vary in intensity of feeling on that dimension. Once the adjectives have been identified and their discriminating potential demonstrated, the selected items are utilized to measure the feelings (attitudes or values) that individual subjects attach to the object.

Typical instructions to subjects are as follows:

Directions: This instrument is designed to measure the meaning of certain things by having people judge them with a series of scales using word opposites. Make your judgments on the basis of what these things mean to YOU. Below you will see the thing to be judged in the center of the page. You are to rate this object on each of the scales below the object. Here is how you use the scales:

If you feel the object in the center is very closely related to one end of the scale, you should place your check-mark as follows:

```
        fair__X__|_____|_____|_____|_____|_____|_____unfair

                                or

        fair_____|_____|_____|_____|_____|_____|__X__unfair
```

If you feel the concept is quite closely related to one or the other end of the scale (but not extremely), you should place your check-mark as follows:

```
        strong_____|__X__|_____|_____|_____|_____|_____weak

                                or

        strong_____|_____|_____|_____|_____|__X__|_____weak
```

If the object seems only slightly related to one side as opposed to the other side (but is really not neutral), then you should check as follows:

```
        active_____|_____|__X__|_____|_____|_____|_____passive

                                or

        active_____|_____|_____|_____|__X__|_____|_____passive
```

If you consider the concept to be neutral on the scale, both sides equally associated with the object, or if the scale is completely irrelevant, unrelated to the concept, then you should place your check-mark in the middle space:

```
        safe_____|_____|_____|__X__|_____|_____|_____dangerous
```

GO AHEAD!

SCHOOL

1. good _____|_____|_____|_____|_____|_____|_____ bad
2. kind _____|_____|_____|_____|_____|_____|_____ cruel
3. high _____|_____|_____|_____|_____|_____|_____ low
4. hard _____|_____|_____|_____|_____|_____|_____ soft
5. heavy _____|_____|_____|_____|_____|_____|_____ light

Typically, 3 or more items are selected from those items which "load" heaviest on each of the factors or dimensions of the semantic space which the researcher wishes to measure. More items from a given dimension yields a more reliable estimate of that dimension. Note that if items from more than one factor are used, a profile of scores may be obtained for each individual. The user of the semantic differential scales may choose, of course, to measure on only one dimension. Items may also be included that are not previously known to load on a particular dimension but are felt by the test constructor to be relevant for measureing the meaning or attitude toward a given object. Later analyses may then be performed to determine the extent to which these other items load on the dimensions of the semantic space.

While it is assumed that the scales (items) of the semantic differential scales are equal interval scales, this assumption may be checked by using the successive interval scaling program to estimate the interval widths of the individual items. Dimension scores for individuals are usually computed by simply summing or averaging the scale values of each item where the scale values are $-3, -2, -1, 0, +1, +2$ and $+3$ corresponding to the seven categories used. Notice that the values may need to be reversed if the "negative" synonym is listed first and the "positive" listed last.

Behavior Checklists

The industrial technology evaluator will sometimes utilize a behavior checklist form to record observations regarding work habits, verbal interactions, or events considered important to a given study. In industrial training situations, the evaluator may record such details as the number of steps taken during a given operation, the frequency of lifting objects from below waist level, the number of manual adjustments to equipment, etc. related to the training. Time and motion studies may provide valuable information for reducing fatigue and injury, reducing operating times for processes, and suggest alternative methods of operation. In evaluating trainer performance, a behavior checklist may "zero in" on specific behaviors potentially detracting from the effectiveness of the instructor as well as identifying those important to retain and reinforce.

As an example of a behavioral checklist, consider the following set of "items" by which trainees record their observations about behaviors of a trainer:

Behavior of the Trainer

Directions: Each item below describes a behavior that you might have observed during the training session. For each item indicate whether or not the behavior occurred and indicate how you felt about the behavior. Express your feeling about the behavior by checking one of the numbers between 1 and 5 where 1 indicates "Highly undesirable", 2 indicates slightly undesirable, 3 indicates neither desirable or undesirable, 4 indicates somewhat desirable and 5 indicates "Definitely desirable".

	ITEM	OBSERVED? (Y OR N)	FEELING 1 2 3 4 5
1.	Embarrassed a trainee.	_____	__ __ __ __ __
2.	Arrived late for a session.	_____	__ __ __ __ __
3.	Showed enthusiasm for the subject.	_____	__ __ __ __ __
4.	Showed a good sense of humor.	_____	__ __ __ __ __
5.	Showed sensitivity to the learner.	_____	__ __ __ __ __
6.	Got off the subject.	_____	__ __ __ __ __
7.	Talked over my head.	_____	__ __ __ __ __
8.	Reviewed what we had learned.	_____	__ __ __ __ __
9.	Handed out helpful reading material.	_____	__ __ __ __ __
10.	Used inappropriate English.	_____	__ __ __ __ __

To "score" the above type of data, the evaluator may multiply the value of the "feeling" scale checked by one (1) if the observer marked "y" to observing it or zero (0) if not observed. The higher the score, the "better" the trainer behaved in the view of the trainees.

Codifying Personal Interactions

In some situations, it is necessary to evaluate the content of interpersonal communications. For example, to create a work environment free of discrimination, the conversations among employees may be coded for words, phrases, sentences, gestures, or behaviors which may be construed as sexist, discriminatory or derogatory to other individuals. Unfortunately, one cannot always sit and take notes while others are conversing. Use of tape recording without the permission of those recorded is also inappropriate. Often the best one can do is to take note of a part of a conversation overheard, record one's observations as soon as possible afterwards, and then, if possible, verify what was heard with one or more persons that may also have heard the conversation. Clearly, this is an emotionally laden and sensitive area! One must use extremely good judgment. Rather than recording specific "offenders" names, for example, one may use code letters or numbers to represent individuals. One may also encode words, gestures, etc. within categories. Let's consider an example where a female employee has complained of sexual harassment in a business which employs primarily men and very few women in packaging meat for retail store distribution. A consultant is hired to evaluate the work place for evidence of a problem with sexual harassment. The evaluator first does a "walk-through" to garner any graphical evidence of harassment such as :

g1 = sexually explicit graffiti or pictures in view in restrooms
g2 = written material making explicit sexual innuendoes regarding an employee

Next, the evaluator may draw a random sample of employees and formally interview them, giving full assurance of confidentiality. The evaluator may code each employees responses as E1, E2, etc. and, using a pre-defined schedule of questions, code the responses to each question as + or − to indicate statements made that verify or negate the presence of harassment. Again, the coding for the questions and their responses might be:

$$E1(1)+;\ E1(2)-;\ E1(3)-;E1(4)+$$
$$E2(1)-;E2(2)+;\ E2(3)+;\ E2(4)-$$
$$\text{etc.}$$

The evaluator may specifically interview the females in the work-setting (recognizing that sexual harassment *can* be evidenced by either gender, but more likely reported by females). This type of interview is again, very sensitive. An individual often must show great courage to even raise the complaint of harassment and may fear reprisal from coworkers or employer. The evaluator must be particularly well versed in the separation of *perceptions* of harassment from *evidence* of harassment. Again, coding of responses to questions or volunteered information may be useful for assuring confidentiality and brevity in data collection. Something like the previous coding might be used:

C1(1)V+; C1(2)P-; etc. where C1 is the first complainant, V is evidence, P is a perception and + or − is content within the definition of harassment or not in the definition of harassment.

Once such data is collected and summarized, the evaluator must still attach weight to each type of evidence or perception. Typically, "hard" evidence such as graffiti, derogatory written comments, verified derogatory conversations, etc. are given a higher value than perceptions or hearsay evidence. Notice that the evaluator is not in the role of changing the work environment, filing complaints with the Equal Opportunity Commission or other corrective decisions and actions. The evaluator in this example was likely asked to determine if harassment exists or perhaps the "degree" of harassment that may exist. The report completed may, of course, suggest alternative actions appropriate to the evidence found and conclusions reached by the evaluator. It is the responsibility of the evaluators employer to act on the evaluation results, not the evaluator.

Bibliography

1. Afifi AA, Azen SP. Statistical analysis: a computer oriented approach. New York: Academic Press; 1972.
2. Anderberg MR. Cluster analysis for applications. New York: Academic Press; 1973.
3. Bennett S, Bowers D. An introduction to multivariate techniques for social and behavioral sciences. New York: Wiley; 1977.
4. Besterfield DH. Quality control. 2nd ed. Englewood Ciffs: Prentice-Hall; 1986.
5. Bishop YM, Fienberg SE, Holland PW. Discrete multivariate analysis: theory and practice. Cambridge: The MIT Press; 1975.
6. Blommers PJ, Forsyth RA. Elementary statistical methods in psychology and education. 2nd ed. Boston: Houghton Mifflin Company; 1977.
7. Borg WR, Gall MD. Educational research: an introduction. 5th ed. New York: Longman; 1989.
8. Brockwell PJ, Davis RA. Introduction to time series and forecasting. New York: Springer-Verlag; 1996.
9. Bruning JL, Kintz BL. Computational handbook of statistics. 2nd ed. Glenview: Scott, Foresman and Company; 1977.
10. Campbell DT, Stanley JC. Experimental and quasi-experimental designs for research. Chicago: Rand McNally College Publishing Company; 1963.
11. Chapman DG, Schaufele RA. Elementary probability models and statistical inference. Waltham: Ginn-Blaisdell; 1970.
12. Cody RP, Smith JK. Applied statistics and the SAS programming language, vol. 4. Upper Saddle River: Prentice Hall; 1997.
13. Cohen J, Cohen P. Applied multiple regression/correlation analysis for the behavioral sciences. Hillsdale: Lawrence Erlbaum Associates; 1975.
14. Cohen J. Statistical power analysis for the behavioral sciences. 2nd ed. Hillsdale: Lawrence Erlbaum Associates; 1988.
15. Comrey AL. A first course in factor analysis. New York: Academic; 1973.
16. Cook TD, Campbell DT. Quasi-experimentation: design and analysis issues for field settings. Chicago: Rand McNally College Publishing Company; 1979.
17. Cooley WW, Lohnes PR. Multivariate data analysis. New York: Wiley; 1971.
18. Crocker L, Algina J. Introduction to classical and modern test theory. New York: Holt, Rinehart and Winston; 1986.
19. Diekhoff GM. Basic statistics for the social and behavioral sciences. Upper Sadle River: Prentice Hall; 1996.
20. Edwards AL. Techniques of attitude scale construction. New York: Appleton-Century-Crofts; 1957.

21. Efromovich S. Nonparametric curve estimation: methods, theory, and applications. New York: Springer-Verlag; 1999.
22. Ferrguson GA. Statistical analysis in psychology and education, vol. 2. New York: McGraw-Hill Book Company; 1966.
23. Fienberg SE. The analysis of cross-classified categorical data. 2nd ed. Cambridge: The MIT Press; 1980.
24. Fox J. Multiple and generalized nonparametric regression. Thousand Oaks: Sage Publications; 2000.
25. Freund JE, Walpole RE. Mathematical statistics, vol. 4. Englewood Cliffs: Prentice-Hall; 1987.
26. Fruchter B. Introduction to factor analysis. Princeton: D. Van Nostrand Company; 1954.
27. Gay LR. Educational research: competencies for analysis and application. 4th ed. New York: Macmillan Publishing Company; 1992.
28. Glass GV, Stanley JC. Statistical methods in education and psychology. Englewood Cliffs: Prentice-Hall; 1970.
29. Gottman JM, Leiblum SR. How to do psychotherapy and how to evaluate it: a manual for beginners. New York: Holt, Rinehart and Winston; 1974.
30. Guertin WH, Bailey Jr JP. Introduction to modern factor analysis. Ann Arbor: Edwards Brothers; 1970.
31. Gulliksen H. Theory of mental tests. New York: Wiley; 1950.
32. Hambleton RK, Swaminathan H. Item response theory: principles and applications. Boston: Kluwer-Nijhoff Publishing; 1985.
33. Hansen BL, Chare PM. Quality control and applications. Englewood Cliffs: Prentice-Hall; 1987.
34. Harman HH. Modern factor analysis. Chicago: The University of Chicago Press; 1960.
35. Hays WL. Statistics for psychologists. New York: Holt, Rinehart and Winston; 1963.
36. Heise DR. Causal analysis. New York: Wiley; 1975.
37. Hinkle DE, Wiersma W, Jurs SG. Applied statistics for the behavioral sciences. 2nd ed. Boston: Houghton Mifflin Company; 1988.
38. Huntsberger DH, Billingsley P. Elements of statistical inference. 6th ed. Boston: Allyn and Bacon; 1987.
39. Kelly LG. Handbook of numerical methods and applications. Reading: Addison-Wesley Publishing Company; 1967.
40. Kennedy Jr WJ, Gentle JE. Statistical computing. New York: Marcel Dekker; 1980.
41. Kerlinger FN, Pedhazur EJ. Multiple regression in behavioral research. New York: Holt, Rinehart and Winston; 1973.
42. Lieberman B, editor. Contemporary problems in statistics: a book of readings for the behavioral sciences. New York: Oxford University Press; 1971.
43. Lindgren BW, McElrath GW. Introduction to probability and statistics. 2nd ed. New York: The Macmillan Company; 1966.
44. Marcoulides GA, Schumacker RE, editors. Advanced structural equation modeling: issues and techniques. Mahwah: Lawrence Erlbaum Associates; 1996.
45. Masters T. Practical neural network recipes in C++. San Diego: Morgan Kaufmann; 1993.
46. McNeil K, Newman I, Kelly FJ. Testing research hypotheses with the general linear model. Carbondale: Southern Illinois University Press; 1996.
47. McNemar Q. Psychological statistics, vol. 4. New York: Wiley; 1969.
48. Minium EW. Statistical reasoning in psychology and education, vol. 2. New York: Wiley; 1978.
49. Montgomery DC. Statistical quality control. New York: Wiley; 1985.
50. Mulaik SA. The foundations of factor analysis. New York: McGraw-Hill Book Company; 1972.
51. Myers JL. Fundamentals of experimental design. Boston: Allyn and Bacon; 1966.
52. Nunnally JC. Psychometric theory. New York: McGraw-Hill Book Company; 1967.

53. Olson CL. Essentials of statistics: making sense of data. Boston: Allyn and Bacon; 1987.
54. Payne DA, editor. Curriculum evaluation: commentaries on purpose, process, product. Lexington: D. C. Heath and Company; 1974.
55. Pedhazur EJ. Multiple regression in behavioral research: explanation and prediction, vol. 3. Fort Worth: Holt, Rinehart and Winston; 1997.
56. Press, WH., Flannery, BP., Teukolsky, SA., and Vetterling, WT. Numerical recipes in C. The art of scientific computing. Cambridge: Cambridge University Press, 1988.
57. Ralston A, Wilf HS. Mathematical methods for digital computers. New York: Wiley; 1966.
58. Rao CR. Linear statistical inference and its applications. New York: Wiley; 1965.
59. Rao V, Rao H. C++ neural networks and fuzzy logic, vol. 2. New York: MIS Press; 1995.
60. Rogers J. Object-oriented neural networks in C++. San Diego: Academic Press; 1997.
61. Roscoe JT, Research F. Statistics for the behavioral sciences, vol. 2. New York: Holt, Rinehart and Winston; 1975.
62. Rummel RJ. Applied factor analysis. Evanston: Northwestern University Press; 1970.
63. Scheffe' H. The analysis of variance. New York: Wiley; 1959.
64. Schumacker RE, Lomax RG. A Beginner's guide to structural equation modeling. Mahwah: Lawrence Erlbaum Associates; 1996.
65. Siegel S. Nonparametric statistics for the behavioral sciences. New York: McGraw-Hill Book Company; 1956.
66. Silverman EN, Brody LA. Statistics: a common sense approach. Boston: Prindle, Weber and Schmidt; 1973.
67. SPSS, Inc. SPSS-X User's guide. 3rd ed. Chicago: SPSS; 1988.
68. Steele, SM. Contemporary approaches to program evaluation: implications for evaluating programs for disadvantaged adults. Syracuse: ERIC Clearinghouse on Adult Education (undated).
69. Stevens J. Applied multivariate statistics for the social sciences, vol. 3. Mahwah: Lawrence Erlbaum Associates; 1996.
70. Stodala Q, Stordahl K. Basic educational tests and measurement. Chicago: Science Research Associates; 1967.
71. Thomson G. The factorial analysis of human ability. 5th ed. Boston: Houghton Mifflin Company; 1951.
72. Thorndike RL. Applied psychometrics. Boston: Houghton Mifflin Company; 1982.
73. Thorndike RL, editor. Educational measurement, One dupont circle. 1971st ed. Washington, DC: American Council on Education; 1971.
74. Veldman DJ. Fortran programming for the behavioral sciences. New York: Holt, Rinehart and Winston; 1967. p. 308–17.
75. Walker HM, Lev J. Statistical inference. New York: Henry Holt and Company; 1953.
76. Winer BJ. Statistical principles in experimental design. New York: McGraw-Hill Book Company; 1962.
77. Worthen BR, Sanders JR. Educational evaluation: theory and practice. Belmont: Wadsworth Publishing Company; 1973.
78. Yamane T. Mathematics for economists: an elementary survey. Englewood Cliffs: Prentice-Hall; 1962.

Index

A
Alpha, 29, 33, 34, 38, 51, 107, 109, 111, 140, 157, 169, 181
Analysis of covariance, 144–146
Analysis of variance, 25, 111–157, 161, 162, 164, 192–194, 197
Analysis of variance using multiple regression methods, 136–141
Arithmetic mean, 19–22
Attitudes, values, beliefs, 210–211
Autocorrelation, 76–85

B
Basic statistics, 1–18
Bayesian statistics, 10–15
Behavior checklists, 228–229
Beta, 10, 29, 105, 148, 155, 169
Binomial distribution, 4, 8, 14, 15, 45–46, 159
Both moving average, 77

C
Canonical correlation, 85, 147–152
Central tendency, 19, 43, 54
Chi-squared distribution, 47–48
Classical item analysis, 200–201
Cluster analyses, 153–154
Cochran Q test, 163
Combination, 7–8, 88, 101, 107, 116, 119, 127–130, 134, 142, 144, 153, 166, 171, 185, 189
Completely randomized design, 112–114
Composite test reliability, 189–190
Concurrent validity, 185, 188
Conditional probability, 8–10
Confidence intervals, 39–40, 66, 69, 71
Construct validity, 186–187
Content validity, 187–188
Contingency chi-square, 160
Covariance, 60, 106, 108, 109, 111, 123, 125, 133, 144–146, 181–183, 187
CUSUM chart, 168–169

D
Decision risks, 28–30
Defect (non-conformity) c chart, 170
Defects per unit u chart, 170
Derivative, 13, 64, 94–95, 97–102, 104, 105, 148
Deviation scores, 57, 104, 108, 153
Differences between correlations in dependent samples, 72–74
Differentiating, 56, 95–97, 222
Discriminant function, 152–153
Discriminate validity, 185–186
Dummy coding, 140–141

E
Effect coding, 137–139, 141–143
Eigenvalues and eigenvectors, 148–151
Estimating population parameters, 17, 23–25, 27, 159
Events, 4–10, 15, 17, 27, 46, 154, 164–166, 228

F
Factor analysis, 155–157, 205, 226
Fisher's exact test, 161
Fixed and random effects, 115

F ratio distribution, 48–49
Frequency distributions, 40–42
Friedman two way ANOVA, 164

G
General linear model (GLM), 88, 146–147
Greco-Latin squares, 127–136
Greek symbols, 1, 2, 24, 29
Guttman scalogram analysis, 221–225

H
Hierarchical cluster analysis, 153–154
Hypotheses about correlations in one population, 67–68
Hypotheses related to a single mean, 30–33

I
Interactions, 116, 118–120, 123, 124, 128, 131, 132, 134, 136, 141–145, 191, 193–196, 228–230
Interval scales, 177–179, 214, 228
Item and test analysis, 199–200
Item banking, 209–210
Item difficulty, 195, 201–203, 205–207, 209
Item discrimination, 200–201, 203
Item response theory, 203–204

K
The Kaplan-Meier survival test, 166
Kendall's coefficient of concordance, 161
Kendall's tau and partial tau, 166
Kolmogorov-Smirnov test, 166
Kruskal-Wallis one-way ANOVA, 161–162
Kuder-Richardson formula, 181–184, 199
Kurtosis, 44–45

L
Latin and Greco-Latin square designs, 127–136
Latin square, 128–136
The Law of large numbers, 6–7
Least squares calculus, 90–92
Least-squares fit, 62–65
Likelihood surface, 12, 13, 16
Likert scaling, 225–226
Linear programming, 171–173
Linear regression, 62, 87–90, 147

Local minima, 16–17
Log-likelihood, 15–17

M
Mann-Whitney U test, 160–161
Maximum likelihood, 10–14, 16, 147, 209
Mean, 1, 2, 5, 13, 19–28, 30–40, 42–45, 47, 54, 56, 57, 60, 66, 77, 78, 103, 109, 111, 112, 114, 115, 117, 118, 121, 123, 124, 126, 127, 130, 131, 135–138, 140, 141, 144, 145, 153, 165, 167–170, 177, 178, 180, 189, 190, 194, 198, 199, 201, 202, 207, 208, 210, 213, 214, 217, 219
Median, 5, 21, 43–44, 177, 203, 213, 219, 220
Minimisation, 14
Multiple regression, 85, 87–110, 136–146, 148, 155
Multiple regression coefficient, 106–107
Multiplication rule of probability, 7
Multivariate analysis of variance, 148, 153

N
Nominal scales, 164, 176
Non-parametric statistics, 159–166
Normal distribution, 15, 20, 31, 34–36, 39, 42, 43, 49, 67, 167, 215
Normal equations, 103–105
Null and alternate hypotheses, 37
Null hypothesis, 28, 29, 31–35, 37, 38, 40, 68, 69, 113, 118, 162

O
One between, one repeated design, 121–124
One parameter logistic model, 205–209
Optimisation, 14, 16, 17
Ordinal scales, 163, 176–177
Orthogonal coding, 138–140, 143, 144

P
Parameters, 1, 4, 5, 10–19, 23–25, 27, 42, 77, 106, 115, 124, 132, 159, 164, 173, 202, 203, 205–210
Partial correlation, 74–75, 85, 186
Partial derivatives, 64, 100–102, 104, 105, 148
Path analysis, 154–155
p chart, 169–170

Permutation, 7, 8
Permutations and combinations, 7–8
Poisson distribution, 46–47, 159, 170
Polynomial regression, 79, 81
Power, 7, 33–36, 129, 140, 146, 162, 203
Prediction equation, 65, 88, 104, 107
Predictive validity, 185, 188
Principal components, 156
Probability, 3–13, 15, 17, 28, 31, 33–35, 37, 39–43, 45, 46, 69, 73, 107, 111, 153, 161–165, 167, 169, 203, 205, 206, 219
Probability of a binomial event, 164–165
Product moment correlation, 53–85

R
Random effects, 115, 120, 126
Range chart, 168
Rasch one-parameter logistic model, 205
Ratio scales, 159, 178–179
Reliability, 155, 179–185, 187–190, 194–199, 202, 210, 224
Reliability by ANOVA, 190–199
Runs test, 165

S
Sample size, 24–26, 29–31, 34–39, 49, 69, 121, 133, 165
Sample space, 5, 6, 17
Sample variance, 24, 25
Sampling, 17–18, 20, 21, 28, 45, 49, 68, 111, 112, 155, 157, 159, 161, 162, 164–166, 188, 197, 210
Scales of measurement, 176–179
Scattergrams, 54–57
S control chart, 168
Semantic differential scales, 226–228
Semi-partial correlation, 75–76
Sign test, 163–164
Simplex, 173
Skew, 44
Spearman rank correlation, 160, 161
Standard error of estimate, 66, 67, 107, 108
Standard error of the mean, 25–27, 31, 32, 39, 40
Standardized scores, 105–107
Statistical process control, 167–170
Statistics, 1–51, 61, 67–72, 77, 87, 88, 106–109, 112, 116, 121, 124, 145–148, 153, 154, 159–166, 175, 176, 187, 209, 213

Successive interval scaling procedures, 217–221
Summation operator, 2, 3

T
Tailored testing, 206, 210
Testing equality of correlations in two populations, 70–72
Testing the difference between regression coefficients, 109–110
Testing the regression coefficients, 107–109
Test length, 188–189
Test theory, 175–176, 205
Thurstone paired comparison scaling, 214–217
Two factor ANOVA by multiple regression, 141–144
Two factor repeated measures analysis, 125
Two-way, fixed-effects, 115–120
Type I error, 28–32, 34, 35, 37–39, 111
Type II error, 28–30, 33–36, 38, 39, 169

U
Unbiased estimate, 20, 24–26, 31, 49, 60, 66

V
Validity, 184–189, 210
Variance and standard deviation, 5, 22–23, 25, 27
Variance of errors of prediction, 66–67
Variance of predicted scores, 66

W
Wilcoxon matched-pairs signed ranks test, 162

X
XBAR chart, 167–168

Z
z scores, 31–34, 36, 37, 39, 40, 42, 43, 47–49, 56–61, 63–65, 71, 123, 148, 154–156, 162, 189, 190, 206, 215, 217–220

MIX
Papier aus verantwortungsvollen Quellen
Paper from responsible sources
FSC® C105338

If you have any concerns about our products,
you can contact us on
ProductSafety@springernature.com

In case Publisher is established outside the EU,
the EU authorized representative is:
**Springer Nature Customer Service Center GmbH
Europaplatz 3, 69115 Heidelberg, Germany**

Printed by Libri Plureos GmbH
in Hamburg, Germany